AF598019

Methods in Molecular Biology

Series Editor
John M. Walker
School of Life and Medical Sciences
University of Hertfordshire
Hatfield, Hertfordshire, AL10 9AB, UK

For further volumes:
http://www.springer.com/series/7651

Kinase Screening and Profiling

Methods and Protocols

Edited by

Hicham Zegzouti and Said A. Goueli

Promega Corporation, Madison, Wisconsin, USA

Editors
Hicham Zegzouti
Promega Corporation
Madison, WI, USA

Said A. Goueli
Promega Corporation
Madison, WI, USA

ISSN 1064-3745 ISSN 1940-6029 (electronic)
Methods in Molecular Biology
ISBN 978-1-4939-3072-2 ISBN 978-1-4939-3073-9 (eBook)
DOI 10.1007/978-1-4939-3073-9

Library of Congress Control Number: 2015951783

Springer New York Heidelberg Dordrecht London

Printed on acid-free paper

Humana Press is a brand of Springer
Springer Science+Business Media LLC New York is part of Springer Science+Business Media (www.springer.com)

Preface

Protein kinases play a central role in cellular physiology and diverse biological processes and thus alterations in their expression or activity have considerable consequences. Based on the critical role they play in regulation of normal and abnormal cellular growth and differentiation, these enzymes prove to be very valuable therapeutic targets. It has been estimated that over 60% of current discovery programs in pharmaceutical companies are focused on protein kinase inhibitors. As of 2015 there are 30 small molecule kinase inhibitors approved for human oncology and for autoimmune diseases and over 150 are currently in clinical trials. These drugs have grossed over $30 billion and projected to reach over $40 billion in 2 years. It is worth noting that the majority of these clinical trials target only 42 protein kinases and about 50% of these inhibitors target kinases which already have approved drugs. Furthermore, over 100 kinases have unknown function and 50% of all kinases are largely uncharacterized. Thus there is significant hope for breakthrough therapies directed towards previously untargeted kinases.

Although so much success has been achieved with developing novel therapeutics for protein kinases in multiple disorders, many hurdles and obstacles had to be overcome at the early stages of considering protein kinases as validated drug target. These include the high concentration of cellular ATP for developing ATP competitive inhibitors, structural similarities for developing selective inhibitors against specific kinase, cellular permeability of the drugs, and above all cellular toxicities as kinases are involved in multiple functions and play intricate role in several signaling pathways. Incorrect or incomplete assessment of cellular selectivity substantially increases the risk of unexpected toxicities at later stages of drug development.

In addressing these issues and with the expectations of expanding the therapeutic potential of targeting as many kinases as possible, we believe a volume devoted to the screening and profiling of protein kinase inhibitors or activators will be timely. The volume will cover current technologies that are in practice at various academic and industrial research laboratories. It covers several facets of the drug discovery processes starting with target identification, assay development, and screening chemical libraries for hit identification and lead optimization. The screening strategies have been mainly done through biochemical assays with emerging trend for cellular kinase screening. The profiling part of the volume discusses several strategies and techniques that are required to minimize off-target hits and minimize cellular toxicities that are caused by liability kinases. We have been fortunate to have a diverse list of experienced authors whose main focus is on the kinases as drug target with different strategies in reaching that goal. We hope this volume will benefit scientists and researchers who are interested in this field of drug discovery.

Madison, WI, USA

Hicham Zegzouti
Said A. Goueli

Contents

Contributors

HEE-CHUL AHN • *Department of Pharmacy, Dongguk University-Seoul, Goyang, Gyeonggi, Republic of Korea*

CHRISTOPHE ANTCZAK • *HTS Core Facility, Memorial Sloan Kettering Cancer Center, New York, NY, USA; Center for Proteomic Chemistry, Novartis Institutes for Biomedical Research, Cambridge, MA, USA*

DOUGLAS S. AULD • *Center for Proteomic Chemistry, Novartis Institutes for Biomedical Research, Cambridge, MA, USA*

SILKE M. BAUER • *Department of Pharmaceutical and Medicinal Chemistry, Institute of Pharmacy, Eberhard-Karls-University Tuebingen, Tuebingen, Germany*

EUN JEONG CHO • *Texas Screening Alliance for Cancer Therapeutics, and College of Pharmacy, The University of Texas at Austin, Austin, TX, USA*

KEVIN N. DALBY • *Division of Medicinal Chemistry and Texas Screening Alliance for Cancer Therapeutics, The University of Texas at Austin, Austin, TX, USA*

MINDY I. DAVIS • *Division of Preclinical Innovation, National Center for Advancing Translational Sciences, National Institutes of Health, Rockville, MD, USA*

CHANNING J. DER • *Department of Pharmacology, University of North Carolina at Chapel Hill, Chapel Hill, NC, USA*

ASHWINI K. DEVKOTA • *Texas Screening Alliance for Cancer Therapeutics, The University of Texas at Austin, Austin, TX, USA*

HAKIM DJABALLAH • *HTS Core Facility, Memorial Sloan Kettering Cancer Center, New York, NY, USA; Institut Pasteur Korea, Seongnam-si, Gyeonggi-do, South Korea*

KRISNA C. DUONG-LY • *Cancer Biology Program, Fox Chase Cancer Center, Philadelphia, PA, USA*

LEANNA R. GENTRY • *Department of Pharmacology, University of North Carolina at Chapel Hill, Chapel Hill, NC, USA*

SAID A. GOUELI • *Promega Corporation, Research and Development, Madison, WI, USA; University of Wisconsin School of Medicine and Public Health, Madison, WI, USA*

KLAUS M. HAHN • *Department of Pharmacology, University of North Carolina at Chapel Hill, Chapel Hill, NC, USA*

BYEONGGU HAN • *Department of Pharmacy, Dongguk University-Seoul, Goyang, Gyeonggi, Republic of Korea*

JACQUELYN HENNEK • *Promega Corporation, Research and Development Department, Madison, WI, USA*

JAMES INGLESE • *Division of Preclinical Innovation, National Center for Advancing Translational Sciences, National Institutes of Health, Rockville, MD, USA*

YONG JIA • *Genomics Institute of the Novartis Research Foundation, San Diego, CA, USA*

JOSE JUAREZ • *Genomics Institute of Novartis Research Foundation, San Diego, CA, USA*

ANDREI V. KARGINOV • *Department of Pharmacology, University of Illinois at Chicago, Chicago, IL, USA*

JAKUB M. KUBIAK • *Department of Pharmaceutical and Medicinal Chemistry, Institute of Pharmacy, Eberhard-Karls-University Tuebingen, Tuebingen, Germany*
SHENSHEN LAI • *Kinexus Bioinformatics Corporation, Vancouver, BC, Canada*
STEFAN LAUFER • *Department of Pharmaceutical and Medicinal Chemistry, Institute of Pharmacy, Eberhard-Karls-University Tuebingen, Tuebingen, Germany*
MARI MANUIA • *Genomics Institute of the Novartis Research Foundation, San Diego, CA, USA*
CHAD J. MILLER • *Department of Pharmacology, Yale University School of Medicine, New Haven, CT, USA*
DEHUA PEI • *Department of Chemistry and Biochemistry, The Ohio State University, Columbus, OH, USA*
STEVEN PELECH • *Kinexus Bioinformatics Corporation, Vancouver, BC, Canada; Department of Medicine, University of British Columbia, Vancouver, BC, Canada*
JEFFREY R. PETERSON • *Cancer Biology Program, Fox Chase Cancer Center, Philadelphia, PA, USA*
ULRICH ROTHBAUER • *Department of Pharmaceutical and Medicinal Chemistry, Institute of Pharmacy, Eberhard-Karls-University Tuebingen, Tuebingen, Germany*
MORITOSHI SATO • *Graduate School of Arts and Sciences, The University of Tokyo, Tokyo, Japan*
XIAOQING SHI • *Kinexus Bioinformatics Corporation, Vancouver, BC, Canada*
ANDREW W. TAO • *University of Michigan, Ann Arbor, MI, USA*
CLINT D.J. TAVARES • *Division of Medicinal Chemistry, The University of Texas at Austin, Austin, TX, USA*
THI B. TRINH • *Department of Chemistry and Biochemistry, The Ohio State University, Columbus, OH, USA*
BENJAMIN E. TURK • *Department of Pharmacology, Yale University School of Medicine, New Haven, CT, USA*
YOSHIO UMEZAWA • *Department of Chemistry, School of Science, The University of Tokyo, Tokyo, Japan*
JOLANTA VIDUGIRIENE • *Research and Development Department, Promega Corporation, Madison, WI, USA*
DIRK F.H. WINKLER • *Kinexus Bioinformatics Corporation, Vancouver, BC, Canada*
HICHAM ZEGZOUTI • *Promega Corporation, Research and Development Department, Madison, WI, USA*
HONG ZHANG • *Kinexus Bioinformatics Corporation, Vancouver, BC, Canada*

Chapter 1

HTRF Kinase Assay Development and Methods in Inhibitor Characterization

Yong Jia, Mari Manuia, and Jose Juarez

Abstract

Due to their important roles in cellular signaling and their dysfunctions being linked to diseases, kinases have become a class of proteins being actively pursued as potential drug targets. Biochemical assays for kinases have been developed in various formats to facilitate inhibitor screening and selectivity profiling. Here, we focus on one such technology: homogeneous time-resolved fluorescence (HTRF). In this chapter, we describe the methods of developing an HTRF kinase assay using mutant EGFR enzyme as an example. We show how to determine the kinetic parameter of the enzyme (ATP K_m), as well as how to study the inhibitor mechanism of action (MoA) exemplified by inhibitors of different MoAs. All methods described here can be readily applied to other kinases with minor modifications.

Key words HTRF, TR-FRET, Kinase, Assay development, MoA studies, Inhibitor characterization

1 Introduction

The human kinome constitutes nearly 2 % of the genome and presents one of the largest classes of druggable targets for various diseases [1, 2]. As about half of the intracellular proteins are regulated by phosphorylation-mediated signal transduction, kinases play an important role in numerous aspects of cell biology including proliferation, differentiation, secretion, and apoptosis [1, 3]. The dysfunction of kinases has been implicated in a wide variety of diseases [4–9]. In recent years, several drugs targeting kinases have been successfully launched (Table 1). These target-based drugs have already made a huge impact in the clinic and in improving patient quality of life. The success so far further solidifies the importance of kinases as drug targets.

The continued interest in targeting kinases has prompted the development of numerous kinase assay technologies. These can

Hicham Zegzouti and Said A. Goueli (eds.), *Kinase Screening and Profiling: Methods and Protocols*, Methods in Molecular Biology, vol. 1360, DOI 10.1007/978-1-4939-3073-9_1, © Springer Science+Business Media New York 2016

Table 1
FDA approved small-molecule kinase inhibitor drugs

Name	Target	Company	FDA approval
Afatinib	EGFR/HER2	Boehringer Ingelheim	2013, Non-small-cell lung cancer (NSCLC)
Axitinib	VEGFR1,2,3/ PDGFR/cKIT	Pfizer	2012, Renal cell carcinoma (RCC)
Bosutinib	Bcr-Abl/Src	Pfizer	2012, Chronic myelogenous leukemia (CML)
Crizotinib	ALK/Met	Pfizer	2011, NSCLC (with ALK mutation)
Dasatinib	Multiple targets	BMS	2006, CML, ALL
Erlotinib	EGFR	Genentech	2005, NSCLC
Gefitinib	EGFR	AstraZeneca	2003, NSCLC
Imatinib	Bcr-Abl	Novartis	2001, CML; 2002, GIST
Lapatinib	EGFR/HER2	GSK	2007, HER2+ Breast
Nilotinib	Bcr-Abl	Novartis	2007
Pazopanib	VEGFR2/PDGFR/cKIT	GSK	2009, RCC
Ruxolitinib	JAK	Incyte/Novartis	2011, Myelofibrosis
Sorafenib	Multiple targets	Onyx/Bayer	2005, Renal cancer; 2007, HCC
Sunitinib	Multiple targets	SUGEN/Pfizer	2006, RCC and GIST
Vemurafenib	BRAF	Plexxikon/Genetech	2011, Melanoma

be categorized into four groups: (1) radiometric assays; (2) phospho-antibody-dependent fluorescent/luminescent assays; (3) phospho-antibody-independent fluorescent/luminescent assays; (4) label-free assays. These technologies have been extensively reviewed in the literature [10–16]. Here, we focus on one of the technologies, namely homogeneous time-resolved fluorescence (HTRF).

HTRF® is a trademark of Cis-Bio International (Bagnol/Ceze Cedex, France) and HTRF assays are homogeneous time-resolved assays that generate a signal by fluorescence resonance energy transfer (FRET) between donor and acceptor molecules. The donor is a Eu^{3+} ion caged in a polycyclic cryptate (Eu-cryptate), while the acceptor is a modified allophycocyanin protein. Laser excitation of the donor at 337 nm results in the transfer of energy to the acceptor at 620 nm when they are in close proximity ($\leq$90 Å) leading to the emission of light at 665 nm over a pro-

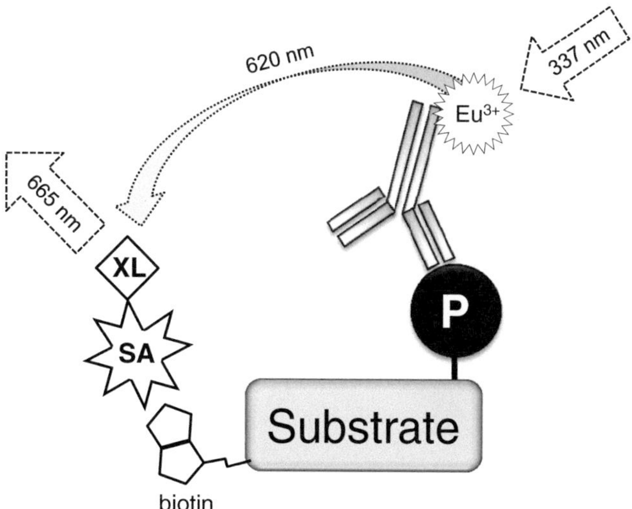

Fig. 1 Principle of HTRF kinase assays: The substrate (either peptide or protein) is biotinylated, which binds to the acceptor SAXL. The donor Eu-cryptate is conjugated to the phospho-specific antibody, which recognizes the generated phosphorylation site on the substrate. The excitation of the donor molecule at 337 nm results in the fluorescence resonance energy transfer to the acceptor molecule at 620 nm, leading to emission of the acceptor molecule at 665 nm. The ratio of 665/620 nm multiplied by 10,000 is recorded as the HTRF signal

longed period of milliseconds. A 50-μs time delay in recording emissions and analysis of the ratio of the 665/620 nm minimizes interfering fluorescence from media components and unpaired fluorophores [17–19]. When formatted for kinase assays, the Eu-cryptate is commonly conjugated to a phospho-specific antibody which binds to the phosphorylated product, while the streptavidin-conjugated allophycocyanin (SAXL) binds to the biotin conjugated to the substrate to complete the detection complex (Fig. 1). HTRF kinase assays are homogeneous, sensitive, versatile, reproducible, environmentally safe, and robust and have been gaining in popularity.

Here, we use one kinase, EGFR(L858R/T790M), as an example to demonstrate the principle of the HTRF assay design, methods in assay development, enzyme kinetic parameter determination, and inhibitor characterization. Mutated EGFR(L858R/T790M) is of major clinical relevance due to its resistance to the approved EGFR TKIs (tyrosine kinase inhibitors) erlotinib and gefitinib [20, 21]. A strong medical need still exists for better treatment of T790M-resistant patients. The methods we describe here for EGFR(L858R/T790M) can be applied to other kinases with minor adaptations.

2 Materials

Prepare all solutions using ultrapure water (prepared by purifying deionized water to attain a sensitivity of 18.2 MΩ cm at 25 °C) and analytical grade reagents. Prepare and store all reagents as indicated. Diligently follow all waste disposal regulations when disposing waste materials. All graph analysis in this chapter were done by GraphPad (San Diego).

2.1 Kinase Reaction

1. 10 % BSA: 5 mg BSA in 50 ml of water (*see* **Note 1**).
2. 1 M tris(2-carboxyethyl)phosphine (TCEP): 2.87 g TCEP in 10 ml of water (*see* **Note 2**).
3. 100 mM Na_3VO_4 (sodium orthovanadate): 183.91 mg Na_3VO_4 in 10 ml of water (*see* **Note 3**).
4. 10× Kinase reaction buffer+: 500 mM HEPES, 100 mM $MgCl_2$, 2 mM $MnCl_2$, adjusted to pH 7.1 with 1 M NaOH (*see* **Note 4**).
5. Kinase reaction buffer: 1× Kinase reaction buffer+, 1 mM TCEP, 0.1 mM Na_3VO_4, and 0.01 % BSA (*see* **Note 5**).
6. 1 mM Substrate-peptide (biotin-(Ahx)-GAEEEIYAAFFA-OH): Substrate peptide was custom synthesized by New England Peptide and was brought to 1 mM with DMSO. Aliquots were stored at −20 °C.
7. 100 mM ATP: Stored at −80 °C.
8. EGFR(L858R/T790M) enzyme, residue 696-1022 (L858R, T790M), 1.8 mg/ml: Expressed and purified as described previously [22].
9. Quenching reagent: 200 mM EDTA. A 0.5 M EDTA solution was diluted to 200 mM with water and stored at room temperature (RT).

2.2 Detection

1. 10 % Tween 20 in water (*see* **Note 6**).
2. Revelation buffer+: 50 mM HEPES, pH 7.0, 400 mM KF, pH adjusted to 7.0 (*see* **Note 7**).
3. Revelation buffer: Revelation buffer+ with 0.01 % Tween 20 and 0.1 % BSA (*see* **Note 8**).
4. SA-XL, "Phycolink™ Streptavidin-$_{XL}$Allophycocyanin Conjugate" (Prozyme): 2 mg/ml in buffer containing 10 mM Tris–HCl, pH 8.2, 150 mM NaCl. Aliquots were stored at 4 °C.
5. PT66-K, Anti-Phosphotyrosine (PT66)-Cryptate (61T66KLB, Lot49A, Cisbio): PT66-K was prepared by reconstituting the

contents of the vial with 1 ml of "revelation buffer+" and 10 μl of BSA to a final concentration of 50 μg/ml and stored at 4 °C.

6. Detection solution: 0.01 μg/μl SA-XL and 0.5 ng/μl PT66-K in revelation buffer (*see* **Note 9**).

2.3 Consumables, Equipment

1. 384-Shallow well solid white Proxiplate (6008289, Perkin Elmer).
2. 384-Well clear round-bottom plate, 120 μl max volume per well (Matrix #4340).
3. 384-Well v-bottom plate (781280, Greiner).
4. Envision HTRF-capable plate reader (Perkin Elmer Life Sciences).
5. MiniTrak equipped with 384 pintool (Perkin Elmer Life Sciences): Used for transferring 50 nanoliter volumes of test compounds to reaction mixtures.
6. 16-Channel matrix electronic multichannel pipette, 2–125 μl: Used for distributing assay reagents into 384-shallow well Proxiplates.

3 Methods

3.1 Enzyme Titration

To develop an enzymatic assay, one first needs to identify the buffer conditions, appropriate substrate, reaction temperature, etc. This information is usually found in published literature; however, it can also be explored/optimized experimentally during the assay development. The assay conditions are likely different for each enzyme, but it is not the intent of this manuscript to cover this topic. Once the initial conditions are identified, the first recommended experiment is to perform an enzyme titration in order to define the amount of enzyme needed for subsequent assays.

1. Thaw enzyme, ATP, and substrate-peptide on ice. Thaw 10× "kinase reaction buffer+," 10 % BSA, 100 mM Na_3VO_4, and 1 M TCEP on bench top and equilibrate to RT.
2. Prepare 10 ml kinase reaction buffer.
3. Prepare ATP/substrate-peptide stock (2 mM ATP, 2 μM substrate-peptide). Add 10 μl of 100 mM ATP stock and 1 μl of 1 mM substrate-peptide to 489 μl of kinase reaction buffer.
4. In a 384-well clear round-bottom plate, add 40 μl of kinase reaction buffer to wells A2 through A12.
5. Dilute the original enzyme stock (1.8 mg/ml) 3600-fold with kinase reaction buffer to 0.5 ng/μl. Add 80 μl of the diluted enzyme in well A1 (Fig. 2 *see* **Note 10**).

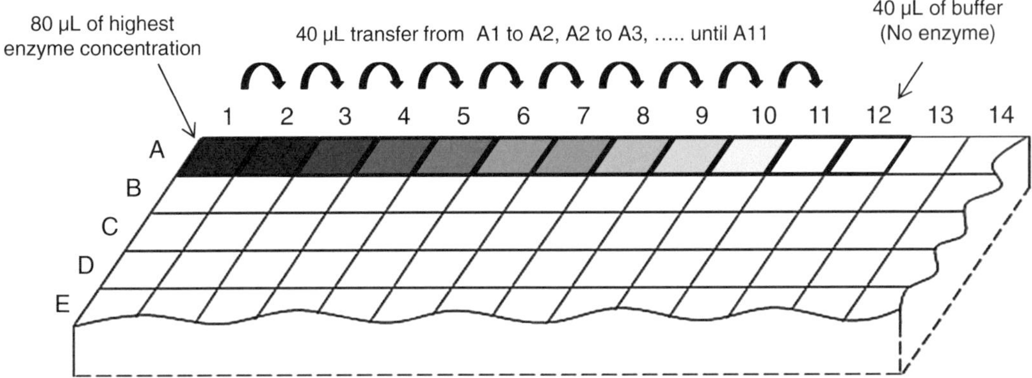

Fig. 2 Enzyme dilution scheme: The upper left quadrant of a 384-well plate is shown. Enzyme stock (80 μl) is added to well A1 and 40 μl of kinase reaction buffer is added to wells A2 through A12. Prepare 11-point serial dilutions of the enzyme by mixing equal volumes of kinase reaction buffer and enzyme stock. Transfer 40 μl from A1 to A2, and mix well by pipetting. Then, transfer 40 μl from A2 to A3, and mix well by pipetting. Repeat the process until A11. Well A12 will contain only kinase reaction buffer without enzyme to serve as negative control

6. Prepare 11-point serial dilutions of the enzyme by mixing equal volumes of kinase reaction buffer and enzyme stock (Fig. 2). Transfer 40 μl from A1 to A2 and mix well by pipetting. Then transfer 40 μl from A2 to A3 and mix well by pipetting. Repeat the process until A11. Well A12 contains only kinase reaction buffer without any enzyme to serve as a negative control.
7. In a 384-shallow well white Proxiplate, distribute 5 μl of ATP/substrate stock to wells A1 through A12 and B1 through B12. The final ATP and substrate-peptide concentrations in the reaction are 1 mM and 1 μM, respectively (*see* **Note 11**).
8. To initiate the reaction, distribute 5 μl of enzyme dilutions (from **step 6**) to each well containing substrate. The final enzyme concentrations in the reaction range from 0 to 2.5 ng/well.
9. Tap the sides of the plate gently to help mix the reagents in the well.
10. Incubate for 60 min at RT (*see* **Note 12**).
11. Quench the reaction by adding 5 μl quenching reagent to all reaction wells. Tap the sides of the plate gently again to help with mixing.
12. Add 5 μl of detection solution to each well. Final concentrations of SA-XL and PT66-K are 0.05 μg/well and 2.5 ng/well, respectively. Incubate at RT in the dark (protected from light) for 60 min.

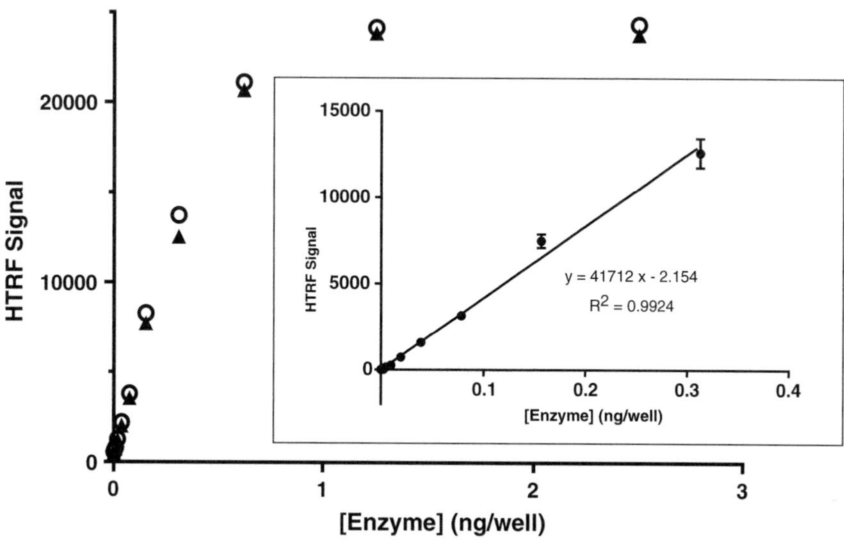

Fig. 3 Enzyme titration curve for EGFR(L858R/T790M)-catalyzed reaction: Duplicate values are shown. Based on the curve, the linear range of the enzyme concentrations is defined as 0 to 0.3 ng/well (inset graph)

13. Read plate on EnVision, with excitation at 337 nm, and dual emission at 665/620 nm. The ratio of 665/620 nm is calculated and reported as final HTRF signal (*see* **Note 13**).
14. Graph the HTRF signal versus the enzyme concentration. A typical titration curve is shown in Fig. 3. Based on the curve, the linear range of the enzyme concentration is defined as 0 to 0.3 ng/well (Fig. 3, inset graph). The concentration of 0.25 ng/well (0.68 nM) enzyme is chosen for further assay development as it is within linear range of enzyme activity and yields excellent S/B (signal/background) window of ~25.

3.2 Enzyme Time Course

After defining the amount of enzyme to be used in the assay from the enzyme titration experiment, the next step is to perform a time course to define the linear kinetic range. This time course will define the duration of the reaction for an end point assay.

1. Prepare the kinase reaction buffer and ATP/substrate-peptide stock the same way as described in Section "Enzyme Titration", **steps 1–3**.
2. Prepare enzyme stock (0.05 ng/μl). Dilute the original enzyme stock (1.8 mg/ml) 100-fold using kinase reaction buffer to 18 μg/ml, and then add 1.39 μl of this diluted enzyme to 499 μl of kinase reaction buffer.
3. Distribute 5 μl of ATP/substrate stock to wells A1 through A12 and B1 through B12 of a 384-shallow-well white Proxiplate. The final ATP and substrate-peptide concentrations in the reaction are 1 mM and 1 μM, respectively.

4. To wells A1 and B1, add 5 μl quenching reagent (*see* **Note 14**).
5. To start the reaction, distribute 5 μl of enzyme stock to all reaction wells (distribute the enzyme stock to wells A1 and B1 last; *see* **Note 15**). The final enzyme concentration in the reaction is 0.25 ng/well (0.68 nM).
6. Tap the sides of the plate gently to help mix the reagents in the well.
7. At designated time points, quench the reaction by adding 5 μl of quenching reagent to wells. In this example, the time points chosen are 0, 5, 10, 15, 20, 25, 30, 40, 50, 60, 70, and 80 min (*see* **Note 16**).
8. Add detection solution and read plate on EnVison as described in Section "Enzyme Titration", **steps 11–12**.
9. Graph the HTRF signal versus time. A typical time course is shown in Fig. 4. From the time course, the linear kinetic ranges from 0 to 80 min. For an end point assay, 60-min duration is chosen as reaction time because it is within the linear kinetic range and provides excellent S/B window (~25).

3.3 ATP K_m Determination

Among all of the kinetic parameters, ATP K_m is perhaps the most important and relevant to drug discovery as most kinase inhibitors are competing with ATP for binding to the kinase. By determining the ATP K_m, one can design experiments to evaluate compound binding mode, i.e., ATP competitive, uncompetitive, or noncompetitive. Using HTRF assay to determine ATP K_m is rather straight-

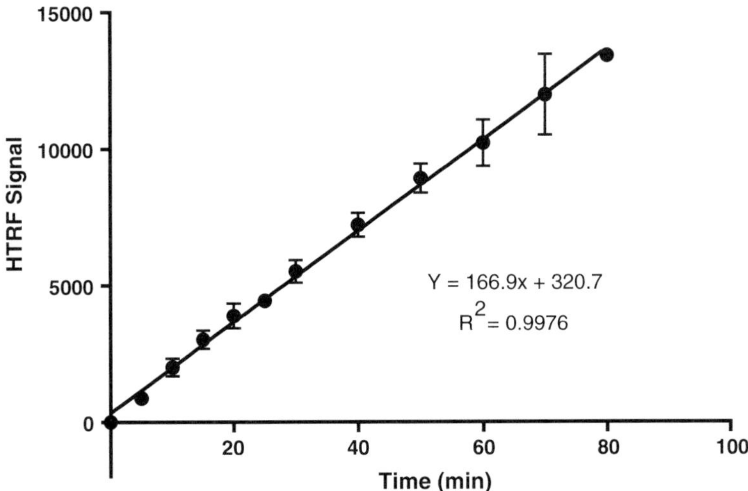

Fig. 4 Time course of EGFR(L858R/T790M)-catalyzed reaction: The linear kinetic ranges from 0 to 80 min. For the end point assay, 60-min reaction time is chosen because it is within the linear kinetic range and provides excellent S/B window of ~25

forward. The initial velocities from each reaction at different ATP concentrations are compared directly, and ATP K_m can be obtained by fitting the data to the Michaelis–Menten equation, Eq. 1 [23], where $v_0$is the initial velocity, V_{max} is the maximal velocity, and K_m is Michaelis constant for ATP. Here we describe the methods for determining the ATP K_m:

$$v_0 = \frac{V_{max}[ATP]}{[ATP] + K_m} \tag{1}$$

1. Prepare enzyme/substrate-peptide stock (0.06 ng/μl enzyme, 2 μM substrate-peptide). Dilute the original enzyme stock (1.8 mg/ml) 100-fold using kinase reaction buffer to 18 μg/ml, and then add 16.8 μl of this diluted enzyme to 5 ml of kinase reaction buffer. To the same tube, also add 10 μl of 1 mM substrate-peptide (*see* **Note 17**).
2. The following various ATP concentration stocks are prepared: 200, 150, 100, 75, 50, 25, 18.8, 12.6, 6.2, 3.2, 1.6, 0.8, and 0.2 μM. To make each ATP stock, the appropriate amount of ATP is added to kinase reaction buffer in a final volume of 1 ml. For example, to make 200 μM ATP stock, add 20 μl of 10 mM ATP to 1 ml of kinase reaction buffer; to make 150 μM ATP stock, add 15 μl of 10 mM ATP to 1 ml of kinase reaction buffer; and so on.
3. Distribute 5 μl of each ATP stock to appropriate wells in a 384-shallow-well white Proxiplate. For example, add 200 μM ATP stock to wells A1 through A17 and B1 through B17, add 150 μM ATP stock to wells C1 through C17 and D1 through D17, and so on. The final ATP concentrations in the reactions will be 100, 75, 50, 37.5, 25, 12.5, 9.4, 6.3, 3.1, 1.6, 0.8, 0.4, and 0.1 μM, respectively.
4. To the wells in Column 1 (wells A1, B1, C1, ... etc), add quenching reagent (*see* **Note 14**).
5. To initiate the reaction, distribute 5 μl of enzyme/substrate-peptide stock to all wells, keeping in mind that wells in Column 1 should be added last (*see* **Note 15**). The final enzyme concentration in the reaction is 0.3 ng/well (0.82 nM), and final substrate-peptide concentration is 1 μM.
6. Tap the sides of the plate gently to help mix the reagents in the well.
7. At designated time points, quench each reaction by adding 5 μl of quenching reagent to wells. In this case, the time points chosen are 0, 2, 4, 6, 8, 10, 12, 14, 16, 18, 20, 22, 24, 26, 28, 30, and 32 min (*see* **Note 18**).
8. Add detection solution and read plate on EnVison as described in Section "Enzyme Titration", **steps 11–12**.

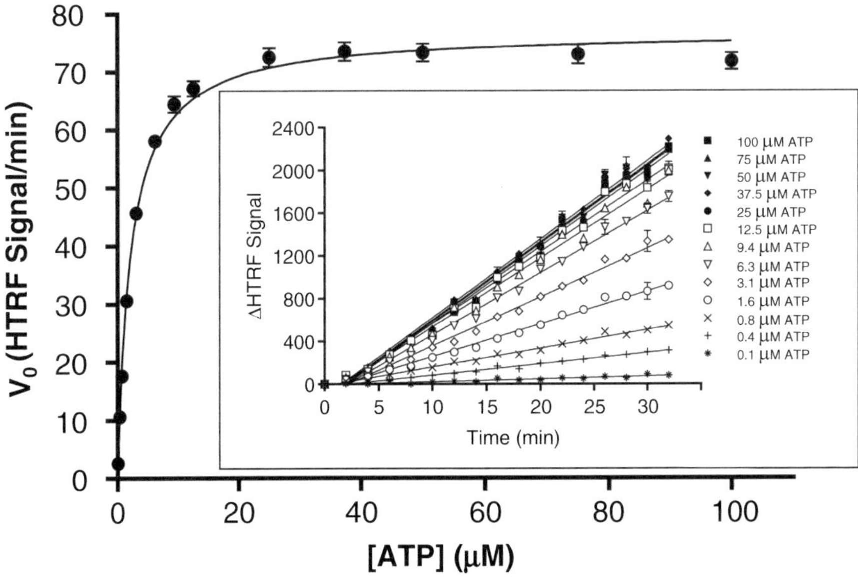

Fig. 5 ATP K_m determination of EGFR(L858R/T790M)-catalyzed reaction: The rates of each reaction are determined by the slopes of each linear time course (inset graph) and the rates versus the ATP concentrations are graphed. The graph is further fitted with the Michaelis–Menten equation, and the apparent ATP K_m is derived to be 2.2 μM. Permissions: Figure 5 is reproduced from Ref. [24] [(Hong L., Quinn C.M., Jia Y. (2009) Evaluating the utility of the HTRF Transcreener™ ADP assay technology: A comparison with the standard HTRF assay technology. Anal Biochem 391, 31–38) by permission of Elsevier Limited (The Boulevard, Langford Lane, Kidlington, Oxford, OX5 1GB, UK). License number 3294871481594; License date Dec 23, 2013]

9. Graph the HTRF signal versus the time for each ATP concentration. A typical graph is shown in Fig. 5, inset. From each time course, the slope of the linear kinetic range is recorded as relative rate. These relative rates are then graphed versus the respective ATP concentrations (Fig. 5), and fitted with the Michaelis–Menten equation, Eq. 1. The derived apparent ATP K_m is 2.2 μM [23] (*see* **Note 19**).

3.4 Inhibitor IC_{50} Determination and Mechanism of Action Studies

The ultimate purpose of assay development is to characterize the inhibitor mode of action which includes determining inhibitor potency (IC_{50}) as well as mechanism of action. IC_{50} determination is carried out by measuring the activity of the enzyme at various inhibitor concentrations, and the inhibitor concentration where 50 % of the enzyme activity is inhibited is defined as IC_{50}. The IC_{50}value thus reflects the compound inhibition potency on a target and is a very important parameter in drug discovery. By comparing the IC_{50} values, compounds can be ranked and lead compounds can be advanced. However, IC_{50} is not the sole reason for compound advancement, as compound MoA is often taken into consideration as well. Kinase inhibitors can be classified into three categories based on their binding mode: (1) compounds binding to ATP site (Type I); (2) compounds binding to both ATP site and the adjacent hydrophobic pocket (Type II); and (3) compounds binding to allosteric sites (Type III). ATP competition experiments can usually differentiate the inhibitor binding modes as follows: Type I inhibitors are fully competitive with ATP, Type II inhibitors compete with ATP to some extent, while Type III inhibitors do not compete with ATP at all (unless compound binding to the allosteric site cross-talks with ATP site and alters ATP binding affinity, which is possible but unlikely) [23]. To further differentiate Type I and Type II inhibitors, time-dependent inhibition experiments can be conducted. Because of the unique binding mode of Type II inhibitors, enzyme conformational change usually accompanies the inhibitor binding resulting in a slow on and slow off rate. ATP competition experiments are done by determining the compound IC_{50} values at various ATP concentrations, and the compound ATP competitiveness is evaluated using Eq. 2 where K_i is the compound inhibition constant and K_m is the Michaels constant for ATP [23]. It is critical to choose the appropriate ATP concentration range to evaluate the compounds. Based on Eq. 2, for ATP-competitive compound, when [ATP] is at its K_m, the IC_{50} equals $2K_i$. When [ATP] is $<K_m$, the resulting IC_{50} will be $<2K_i$. Practically, it is almost impossible to compare different IC_{50}values in the range between $<2K_i$ and $2K_i$ with any confidence as this variation usually lies within the error of the assay technology. Only when the [ATP] is $>>K_m$ do the IC_{50} value increases become more pronounced. For example, when [ATP] = $100K_m$ and $1000K_m$, the IC_{50} is calculated to be ~$100Ki$ and ~$1000K_i$, respectively. We routinely choose a few ATP concentrations ranging from ~K_m to ~$1000K_m$, which yield expected IC_{50} values from ~$2K_i$ to ~$1000Ki$; this dynamic range is large enough to be evaluated with confidence. In addition, ATP site binders can be reversible or irreversible. Irreversible inhibitors engage in a chemical reaction upon binding which is time dependent. To elucidate Type II inhibitor and irreversible inhibitor mechanisms, we also test ATP competition under conditions with or without enzyme/compound preincubation. In this section, we describe how the IC_{50} and ATP competition experiments are designed and executed and how the results are interpreted.

$$IC_{50} = K_i(1 + \frac{[ATP]}{K_m}) \tag{2}$$

1. Prepare serial-diluted compound stock. In a 384-well v-bottom plate, add 20 μl of DMSO to wells B2 through B12, C2 through C12, and D2 through D12. Add 30 μl of the 10 mM compound stocks to wells B1 (afatinib), C1 (erlotinib) or D1 (GNF822). Prepare a 12-point three-fold serial dilution of the compound by transferring 10 μl from B1 to B2, and mix well

by pipetting. Repeat the process from B2 to B3, B3 to B4, and so forth until B12. Repeat the serial dilutions for compounds in rows C and D (*see* **Note 20**). The diluted compound concentrations are 10000, 3333, 1111, 370, 123, 41, 13.7, 4.6, 1.5, 0.51, 0.17, and 0.056 μM. Add 20 μl of DMSO to Rows A and P (wells 1 through 12).

2. Prepare ATP stocks (2, 20, 200, and 2000 μM ATP). Add 5 and 50 μl of 1 mM ATP to 2.5 ml of kinase reaction buffer to make [ATP] of 2 and 20 μM, respectively. Add 5 and 50 μl of 100 mM ATP to 2.5 ml of kinase reaction buffer to make [ATP] of 200 and 2000 μM, respectively (*see* **Note 21**).
3. Prepare enzyme/substrate-peptide stocks (Stock X is 0.1 ng/μl enzyme, 2 μM substrate-peptide; Stock Y is 0.05 ng/μl enzyme, 2 μM substrate-peptide). Dilute the original enzyme stock (1.8 mg/ml) 100-fold using kinase reaction buffer to 18 μg/ml, and then add 14 μl of this diluted enzyme and 5 μl of 1 mM substrate to 2.5 ml of kinase reaction buffer to make enzyme/substrate peptide Stock X; and add 21 μl of this diluted enzyme and 15 μl of 1 mM substrate to 7.5 ml of kinase reaction buffer to make enzyme/substrate-peptide Stock Y. Stock X is for the final ATP concentration of 1 μM, while Stock Y is for the final ATP concentration of 10, 100, and 1000 μM (*see* **Note 22**).
4. Distribute 5 μl of enzyme/substrate-peptide stocks to Rows A through D (wells 1 through 12) in a 384-shallow-well white Proxiplate (*see* **Note 23**). Distribute 5 μl kinase reaction buffer to the last row (wells P1 through 12) to serve as negative control.
5. Using MiniTrak pintool, spike in 50 nL of the serial-diluted compound stock from the compound dilution plate (From **Step 1**). The final compound concentrations in the reactions are 50, 16.7, 5.6, 1.85, 0.62, 0.2, 0.068, 0.023, 0.0076, 0.0025, 0.00085, and 0.00028 μM.
6. Two enzyme/compound preincubation conditions are carried out: without preincubation and with 90-min preincubation (*see* **Note 24**). For the condition without preincubation, after addition of compounds, immediately start the reaction by distributing the ATP stocks to appropriate wells (2 μM ATP stock to enzyme/substrate-peptide Stock X; 20, 200, and 2000 μM ATP stock to enzyme/substrate-peptide Stock Y) (*see* **Note 25**). For 90-min preincubation condition, after addition of compound, leave the plates at RT for 90 min, and then start the reaction by distributing the ATP stocks to appropriate wells. The final concentrations in each reaction condition are (A) 0.5 ng/well (1.36 nM) enzyme, 1 μM substrate-peptide, and 1 μM ATP; (B) 0.25 ng/well (0.68 nM) enzyme, 1 μM substrate-peptide, and 10 μM ATP; (C) 0.25 ng/well (0.68 nM) enzyme, 1 μM substrate-peptide, and 100 μM ATP; and (D) 0.25 ng/well (0.68 nM) enzyme, 1 μM substrate-peptide, and 1000 μM ATP.

7. Tap the sides of the plates gently to help mix the reagents in the well.
8. After 60-min incubation at RT, quench the reaction by adding 5 μl of quenching reagent.
9. Add detection solution and read plates on EnVison as described in Section "Enzyme Titration", **steps 11–12**.
10. Normalize the HTRF signal to "Percent Activity" based on Eq. 3, where the test compound signal is the HTRF readout in each well, positive control is the averaged signal from Row A (no compound), and negative control is the averaged signal from Row P (no enzyme):

$$\%\text{activity} = \frac{\text{Signal} - \text{NegativeControl}}{\text{PostiveControl} - \text{NegativeControl}} \tag{3}$$

11. Graph the "Percent Activity" versus compound concentrations, and fit the curves with nonlinear regression analysis.
12. Data interpretation: A typical outcome for this experiment is exemplified by IC_{50} curves of afatinib (Fig. 6a, b). The IC_{50} data for afatinib, erlotinib, and GNF822 without and with enzyme/compound preincubation are summarized in Table 2. The relationship between IC_{50} value and [ATP] for each compound is illustrated in Fig. 6c. Figure 6a shows that without preincubation, afatinib IC_{50} curves shift to the right (IC_{50}values are increasing) with increasing [ATP], indicating that afatinib is competing with ATP for binding. However, after the preincubation with enzyme, this IC_{50} curve shift is no longer observed (Fig. 6b). These same data are also illustrated in a different way in Fig. 6c and Table 2. Altogether, these data indicate that afatinib occupies the ATP-binding site, but it is not a classic ATP-competitive inhibitor. Its time-dependent inhibition behavior categorizes afatinib as either Type II or irreversible inhibitor class. Further characterization definitively demonstrates the irreversible mechanism of afatinib on EGFR [25]. Figure 6c and Table 2 show that erlotinib IC_{50} values increase with increasing [ATP], and this increase is similar under both no preincubation and 90-min preincubation conditions. These data indicate that erlotinib is a classic ATP-competitive inhibitor of EGFR (Type I) which agrees with the literature report [26]. On the contrary, the IC_{50} values of GNF822 remain largely unchanged versus ATP concentrations or the preincubation time (Fig. 6c and Table 2) suggesting that GNF822 is an allosteric inhibitor of EGFR (Type III). Indeed, an in-house co-crystal structure of EGFR with GNF822 demonstrated that GNF822 binds to a distant allosteric site on the enzyme (unpublished data).

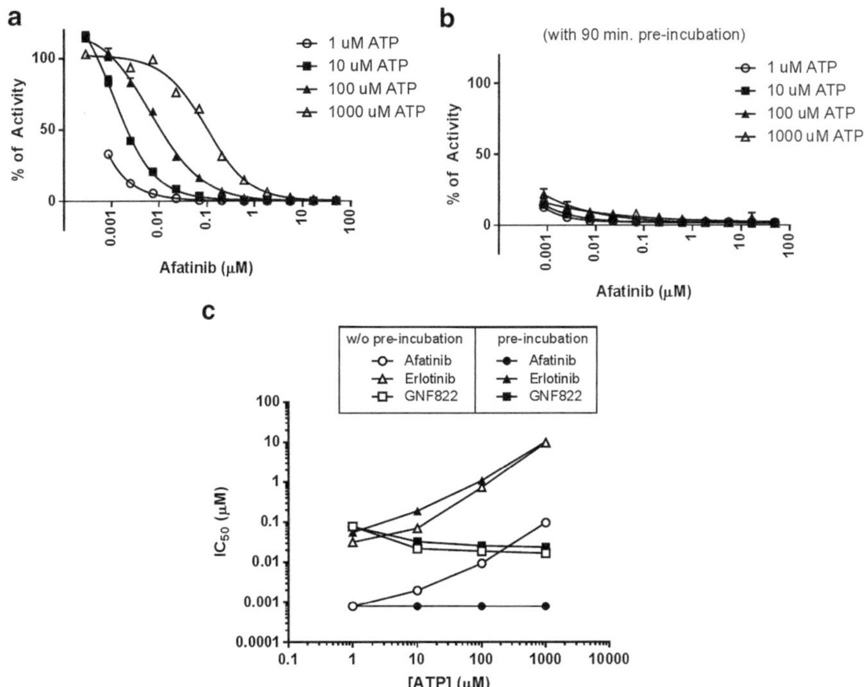

Fig. 6 Inhibitor mechanism of action studies: (**a**, **b**) IC_{50} curves of afatinib on EGFR(L858R/T790M) at different ATP concentrations (**a**) without preincubation and (**b**) with 90-min enzyme/compound preincubation. (**c**) IC_{50} versus [ATP] of afatinib, erlotinib, and GNF822 without and with 90-min enzyme/compound preincubation

Table 2
IC_{50}values of afatinib, erlotinib, and GNF822 on EGFR(L858R/T790M) at various ATP concentrations

	IC_{50} (μM)							
	No preincubation				90-min preincubation			
[ATP] (μM)	1	10	100	1000	1	10	100	1000
Afatinib	<0.0008	0.002	0.0095	0.098	<0.0008	<0.0008	<0.0008	<0.0008
Erlotinib	0.032	0.071	0.74	9.8	0.056	0.19	1.1	10.3
GNF822	0.078	0.022	0.019	0.017	0.077	0.033	0.026	0.024

4 Notes

1. 10 % BSA can be prepared by adding 5 mg to 25 ml of water, and then allowing the BSA to dissolve into the water rotating horizontally on a rotisserie. Once the BSA has completely dissolved, bring the volume up to 50 ml with water, and filter through a 0.22 μm membrane. Aliquot and store at −20 °C.

2. Aliquot 1 M TCEP and store at −20 °C. Upon thawing, TCEP does not always go completely back into solution; warming it in a 37 °C water bath with intermittent mixing will help.
3. Aliquot 100 mM Na_3VO_4 and store at −20 °C. Na_3VO_4 should be activated for maximal inhibition of protein phosphotyrosyl-phosphatases. The following protocol is used to prepare the Na_3VO_4: https://www.wikiformulation.org/protocols/preparation-of-sodium-orthovanadate-solution/10011.
4. 10× Kinase reaction buffer stock without TCEP/Na_3VO_4/BSA can be prepared and stored at −20 °C for long-term use.
5. Prior to use, dilute 10× "kinase reaction buffer+" ten times with water, and add TCEP, Na_3VO_4, and BSA.
6. Due to its very viscous nature, use a plastic, disposable syringe to measure and distribute Tween 20 by pouring into an open syringe to the desired volume and then inserting the plunger. Use a stir bar to aid in the dissolution of Tween 20 into water.
7. Revelation buffer can be prepared without Tween and BSA (named as "revelation buffer +"), filtered through a 0.22 μm filter and stored at 4 °C for long-term storage
8. Prior to using the revelation buffer, *add the following reagents fresh to the revelation buffer+*: 1 μl of 10 % Tween and 10 μl of 10 % BSA per 1 ml.
9. Detection buffer is prepared *immediately prior to use* by adding 5 μl 2 mg/ml SA-XL and 10 μl 50 μg/ml PT66-K into 1 ml revelation buffer.
10. Depending on the enzyme activity, the highest concentration of enzyme used in the titration can vary drastically. A good titration curve will capture both signal escalating phase and saturation phase. Multiple rounds of titrations may need to be performed to better understand the optimal enzyme concentration range. In the case of EGFR(L858R/T790M) reaction, the enzyme is extremely active. Thus, the highest enzyme concentration used in this titration is very low (2.5 ng/well).
11. Based on the experimental design and replicates tested, either the whole or partial plate will be used. For a standard enzyme titration experiment under one substrate condition, we routinely run the assay in duplicates in 12 conditions, which equals to a total of 24 wells. All assays described in this protocol are performed in duplicates.
12. The time point for quenching the reaction is based on the desired duration of the reaction. As the assay is designed for HTS automation, too short of a reaction time is very difficult for the robot handling, while too long of a reaction time is unnecessary and could be problematic for enzyme stability concerns. A 60-min reaction time is generally a good selection for the industry setting.

13. Set the HTRF-capable EnVision reader to HTRF reading mode. The reader utilizes a nitrogen TRF laser that generates an excitation pulse at 337 nm, and the resulting fluorescence is detected at both 665 and 620 nm. An ~50-μs delay in detection was incorporated to reduce background noise introduced by media components and the free acceptor molecules. The final readout is the ratio of 665/620 nm (this ratio is multiplied by 10,000 to make it an integer). This patented 665 nm/620 nm ratio calculation provides real-time quenching correction.
14. The quenching reagent EDTA was added before addition of the enzyme to generate a 0-min time point.
15. It is important to add enzymes to wells A1 and B1 last as these wells contain the quenching reagent EDTA. Otherwise, the tips of the pipettor can be contaminated with EDTA, thus interfering with the reaction of subsequent wells.
16. As the desired duration of the reaction is 60 min, we choose most of the time points before 60 min and a couple of time points beyond 60 min to determine how far the linear kinetics proceeds.
17. In this experiment, a slightly higher concentration of enzyme is used for two reasons: (1) Since much lower concentrations of ATP are used and shorter reaction time is recorded, higher enzyme concentration could compensate for the potential reduction in signal. (2) The kinetics of each reaction is monitored, and only the linear portion of the time course is used to calculate the rate.
18. This experiment measures the rate of each reaction; thus we choose very dense time points up to 30 min which is sufficient to define the linear kinetic range.
19. To derive a reliable ATP K_m, the concentrations of ATP should cover a range of at least from ≥10-fold to ≤10-fold of the K_m value, and ideally with more points around the K_m value. Without knowing the K_m value beforehand, it can be a challenge to design experiments. The usual practice is to choose a wide range of ATP concentrations to start. Based on the derived K_m value (usually a good estimate), further experiments can be fine-tuned. Sometimes multiple rounds of fine-tuning are required to obtain the accurate K_m value. The true ATP K_m is determined under saturating substrate concentration. In this example, the substrate-peptide concentration used in the assay (1 μM) is below its reported K_m (48 μM, [24]); thus the derived ATP K_m from this experiment is considered apparent K_m.
20. The compound dilution of all inhibitors can be carried out simultaneously with a manual multichannel pipettor.

21. Based on the apparent ATP K_m of ~2 μM (Section "ATP K_m Determination"), four ATP concentrations are chosen to evaluate the compound ATP competitiveness: 1, 10, 100, and 1000 μM. According to Eq. 2, if the compounds were ATP competitive, the expected IC_{50}values at these ATP concentrations would be 1.5 *Ki*, 6 *Ki*, 51 *Ki*, and 501 *Ki*, respectively.
22. For each ATP concentration, the enzyme concentration needs to be re-optimized to ensure linearity of the kinetics. This involves the enzyme titration and time course experiments as described in Sections "Enzyme Titration" and "Enzyme Time Course". The initial assay was developed at 1 mM ATP using 0.25 ng/well (0.68 nM) enzyme, and we determined that this enzyme amount also applied to 100 and 10 μM ATP. The optimized enzyme concentration for 1 μM ATP is determined to be 0.5 ng/well (1.36 nM) (the details of the assay re-optimization at different ATP concentrations are not shown).
23. Assays are run in duplicate plates. Four ATP concentration conditions are tested in this experiment, which requires 8 plates total. Enzyme/substrate-peptide Stock X is distributed to plates 1 and 2 (for final [ATP] of 1 μM), Stock Y is distributed to plates 3 and 4 (for final [ATP] of 10 μM), plates 5 and 6 (for final [ATP] of 100 μM), and plates 7 and 8 (for final [ATP] of 1000 μM).
24. A 90-min enzyme/compound preincubation time is chosen here as this time period is generally sufficient to fully capture the events such as inhibitor-induced enzyme conformational change or covalent reaction by irreversible inhibitors. However, this long preincubation time requires the enzyme to be stable at RT for that period of time.
25. Use a 16-channel matrix electronic multichannel pipettor to add ATP from Columns 12 to 1 (from lowest to highest compound concentrations) to avoid compound concentration cross-contamination.

References

1. Manning G, Whyte DB, Martinez R et al (2002) The protein kinase complement of the human genome. Science 298:1912–1934
2. Cohen P (1999) The development and therapeutic potential of protein kinase inhibitors. Curr Opin Chem Biol 3:459–465
3. Hunter T (2000) Signaling—2000 and beyond. Cell 100:113–127
4. Paez JG, Janne PA, Lee JC et al (2004) EGFR mutations in lung cancer: correlation with clinical response to gefitinib therapy. Science 304:1497–1500
5. Pao W, Girard N (2011) New driver mutations in non-small-cell lung cancer. Lancet Oncol 12:175–180
6. Soda M, Choi YL, Enomoto M et al (2007) Identification of the transforming EML4–ALK fusion gene in non-small-cell lung cancer. Nature 448:561–566
7. Davies H, Bignell GR, Cox C et al (2002) Mutations of the BRAF gene in human cancer. Nature 417:949–954
8. Fabbro D, Garcia-Echeverria C (2002) Targeting protein kinases in cancer therapy. Curr Opin Drug Discov Dev 5:701–712

9. Myers MR, He W, Hulme C (1997) Inhibitors of tyrosine kinases involved in inflammation and autoimmune disease. Curr Pharm Des 3:473–502
10. Minor L (2005) Assays for membrane tyrosine kinase receptors: methods for high-throughput screening and utility for diagnostics. Expert Rev Mol Daign 5:561–571
11. Sittampalam GS, Kahl SD, Janzen WP (1997) High-throughput screening: advances in assay technologies. Curr Opin Chem Biol 1:384–391
12. Zaman GJR, Garritsen A, de Boer TH et al (2003) Fluorescence assays for high-throughput screening of protein kinases. Comb Chem High Throughput Screen 6:313–320
13. von Ahsen O, Bömer U (2005) High-throughput screening for kinase inhibitors. Chem Bio Chem 6:481–490
14. Jia Y, Gu X, Brinker A et al (2008) Measuring the tyrosine kinase activity: a review of biochemical and cellular assay technologies. Expert Opin Drug Discov 3:959–978
15. Jia Y, Quinn CM, Kwak S et al (2008) Current in vitro kinase assay technologies: the quest for a universal format. Curr Drug Discov Technol 5:59–69
16. Zegzouti H, Zdanovskaia M, Hsiao K, Goueli SA (2009) ADP-Glo: A bioluminescent and homogeneous ADP monitoring assay for kinases. Assay Drug Dev Technol 7:560–572
17. Mathis G (1999) HTRF technology. J Biomol Screen 4:309–313
18. Mathis G (1995) Probing molecular interactions with homogeneous techniques based on rare earth cryptates and fluorescence energy transfer. Clin Chem 41:1391–1397
19. Mathis G (1993) Rare earth cryptate and homogeneous fluoroimmunoassays with human sera. Clin Chem 39:1953–1959
20. Kobayashi S, Boggon TJ, Dayaram T et al (2005) EGFR mutation and resistance of non-small-cell lung cancer to gefitinib. N Engl J Med 352:786–792
21. Pao W, Miller VA, Politi KA et al (2005) Acquired resistance of lung adenocarcinomas to gefitinib or erlotinib is associated with a second mutation in the EGFR kinase domain. PLoS Med 2, e73
22. Yun C-H, Boggon TJ, Li Y et al (2007) Structure of lung cancer-derived EGFR mutants and inhibitor complexes: mechanism of activation and insights into differential inhibitor sensitivity. Cancer Cell 11:217–227
23. Jia Y, Quinn CM, Gagnon AI (2006) Homogeneous time-resolved fluorescence and its applications for kinase assays in drug discovery. Anal Biochem 356:273–281
24. Hong L, Quinn CM, Jia Y (2009) Evaluating the utility of the HTRF Transcreener™ ADP assay technology: A comparison with the standard HTRF assay technology. Anal Biochem 391:31–38
25. Li D, Ambrogio L, Shimamura T et al (2008) BIBW2992, an irreversible EGFR/HER2 inhibitor highly effective in preclinical lung cancer models. Oncogene 27:4702–4711
26. Moyer JD, Barbacci EG, Iwata KK et al (1997) Induction of apoptosis and cell cycle arrest by CP-358,774, an inhibitor of epidermal growth factor receptor tyrosine kinase. Cancer Res 57:4838–4848

Chapter 2

Application of Eukaryotic Elongation Factor-2 Kinase (eEF-2K) for Cancer Therapy: Expression, Purification, and High-Throughput Inhibitor Screening

Clint D.J. Tavares, Ashwini K. Devkota, Kevin N. Dalby, and Eun Jeong Cho

Abstract

Protein kinases have emerged as an important class of therapeutic targets, as they are known to be involved in pathological pathways linked to numerous human disorders. Major efforts to discover kinase inhibitors in both academia and pharmaceutical companies have centered on the development of robust assays and cost-effective approaches to isolate them. Drug discovery procedures often start with hit identification for lead development, by screening a library of chemicals using an appropriate assay in a high-throughput manner. Considering limitations unique to each assay technique and screening capability, intelligent integration of various assay schemes and level of throughput, in addition to the choice of chemical libraries, is the key to success of this initial step. Here, we describe the purification of the protein kinase, eEF-2K, and the utilization of three biochemical assays in the course of identifying small molecules that block its enzymatic reaction.

Key words Kinase, eEF-2K, Expression, Purification, Fluorescence, Luminescence, Radioactive, High-throughput screen

1 Introduction

Protein kinases catalyze the transfer of phosphate from ATP to target proteins, a process called phosphorylation. Since phosphorylation can modify a protein's function in various ways, kinases play a vital role in the signaling processes within cells. Protein kinases constitute about 2 % of all human genes, with 518 human kinases identified. Eukaryotic elongation factor-2 kinase (eEF-2K), also known as calcium/calmodulin-dependent protein kinase-III (CaM kinase-III), shows increased activity in several cancer cell lines [1–7] and has been identified as a potential therapeutic target for breast cancer, gliomas [1, 2, 8–10], and depression [11]. Recently, Leprivier et al. provided evidence for an important role for eEF-2K

Hicham Zegzouti and Said A. Goueli (eds.), *Kinase Screening and Profiling: Methods and Protocols*, Methods in Molecular Biology, vol. 1360, DOI 10.1007/978-1-4939-3073-9_2, © Springer Science+Business Media New York 2016

in protecting cells from nutrient deprivation, potentially allowing tumors to exploit eEF-2K to support adaptation to metabolic stress [12]. However, recent attempts to identify small-molecule inhibitors of eEF-2K have had mixed results. For example, NH125 has been used in a number of studies to characterize eEF-2K activity in cells [11, 13]. While this histidine kinase inhibitor was reported to inhibit eEF-2K in vitro with an IC_{50} in the nanomolar range [13], our studies suggest that it is a weak promiscuous colloidal aggregator [14]. Moreover, it was found that NH125 could induce the phosphorylation of eEF-2 in mammalian cells [14, 15]. A-484954, a first-generation eEF-2K inhibitor, was recently reported; however it exhibits only moderate potency [15].

Previous assays targeting eEF-2K in drug screening utilized a form of the enzyme possessing a GST-tag, which is inherently dimeric in nature [16] and can interfere with protein structure and function when fused to a protein of interest [17]. Owing to the increasing demand on universal assay strategies against a range of kinases, a large number of reviews have been published [18], and many robust assays have become commercially available. Radioactive-, fluorescence-, and luminescence-based technologies are among the most popular, and have been actively applied to a diverse array of kinases. However, each assay technology possesses its own limitations, and hits must be validated using additional assays.

Radioactive assays quantitating the incorporation of radioactive γ-phosphate from [γ-^{32}P]ATP into a peptide or protein substrate using a scintillation counter is the most traditional method (Fig. 1a). While this method does not interfere with inherent optical properties of compounds often found in chemical libraries, and eliminates nonspecific side reactions, a disadvantage is that it is not practicable for high-throughput applications. Radioactive assays are better suited for confirmation (or orthogonal assays) at lower throughput, or for mechanistic analysis.

Fluorescence or luminescence assays represent advanced options in terms of sensitivity, cost, speed, throughput, and safety. Assays featuring optical signals quantifying either products (e.g., phosphopeptide, ADP) or ATP are now readily accessible from various manufacturers. While luminescence assays have been most broadly commercialized and employed due to their versatility (applicable to any pair of kinases and substrates), the main drawback is the potential interference of compounds with the secondary enzyme (e.g., luciferase) (Fig. 1b). Unlike luminescence assays, fluorescence-based assays offer enhanced flexibility with respect to the choice of fluorophores and assay principles. A robust method is based on the chelation-enhanced fluorophore (CHEF), 8-hydroxy-5-(*N*,*N*-dimethylsulfonamido)-2-methylquinoline (referred to as Sox [19, 20]). Sox is an unnatural amino acid that can be prepared as an Fmoc-protected derivative and incorporated into a substrate peptide (referred to as Soxtide) using standard solid-phase peptide

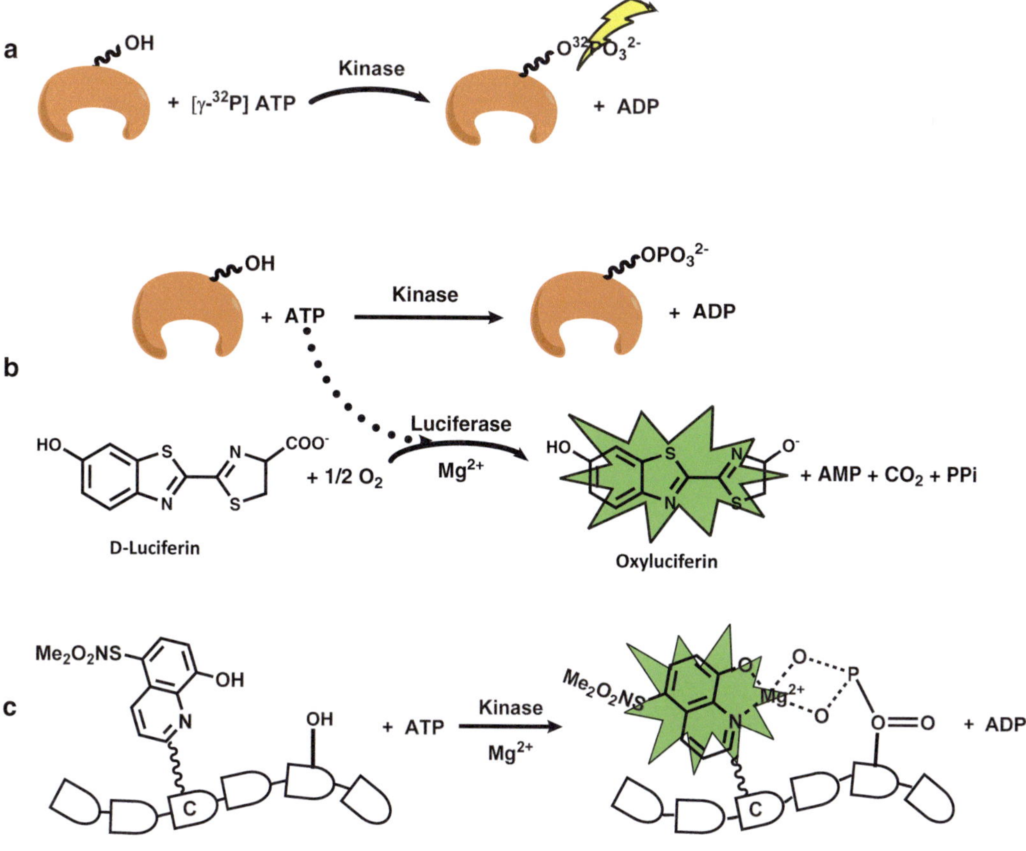

Fig. 1 Schematic models of kinase assays, (**a**) radioactive assay, (**b**) luminescence assay, and (**c**) Sox-based fluorescence assay

chemistry (cysteine alkylation). Upon phosphorylation of the Sox-containing peptide, the binding affinity of the Sox chromophore for Mg^{2+} increases significantly, resulting in a large fluorescence increase (2–12-fold), which can be monitored continuously at 485 nm (Fig. 1c). The commercial availability of assays for 193 different kinases [21] makes this assay useful in high-throughput format, although potential interference from a large fraction of compounds in a typical collection due to autofluorescence is a primary weakness compared to other fluorescence-based methods. As a whole, identification of novel drug candidates is a multi-step endeavor that requires thorough understanding of the capability of individual assay systems and incorporation of available approaches in a proper order to efficiently eliminate any artifacts associated with the particular assay mechanism.

In this regard, we present a series of procedures to screen inhibitors against eEF-2K, covering purification of tag-free target

Table 1
Assay principle and conditions employed in the eEF-2K inhibitor screening

Technology	Detection modality	Substrate	Condition[a]
Fluorescence (FI)	Phosphorylated substrate	F-peptide	384-well plate, 10 μL
			10 nM eEF-2K, 10 μM substrate, 100 μM ATP
Luminescence (Lum)	ATP depletion	Peptide	384-well plate, 10 μL
			5 nM eEF-2K, 30 μM substrate, 1 μM ATP
Radioactive	Phosphorylated substrate	Peptide	96-Well plate, 100 μL
			2 nM eEF-2K, 50 μM substrate, 100 μM [γ-^{32}P]ATP

[a]Assay condition includes (1) assay plate format and volume and (2) composition of key reagents

enzyme, followed by three different assay principles (Table 1). Although we have also characterized this enzyme using detailed kinetic studies, where steady-state analysis on the activity of eEF-2K against a peptide substrate suggests an ordered sequential mechanism with ATP binding first [14, 22], we will not be discussing the kinetics as this is not within the scope of this chapter. However, it is worthwhile to point out that those kinetic studies have shown that this enzyme is suitable for the enzymatic assays that have been employed here. The purified tag-free enzyme is stable at room temperature for over 4 h and is a druggable target. Here, both fluorescence and luminescence assays optimized for 384-well plate formats were found to be robust, automation friendly, speedy, and cost effective. A radiolabeled assay was successfully applied to eliminate false hits post-screening.

2 Materials

2.1 Enzyme Purification and Kinase Assay

1. DH5α strain (Invitrogen Corporation).
2. Rosetta-gami™ 2(DE3) strain (Novagen).
3. pET-32a plasmid (Novagen).
4. *E. coli* culture media: Luria–Bertani (LB) media supplemented with 135 μM ampicillin, 300 μM chloramphenicol, and 20 μM tetracycline.
5. Isopropyl β-d-1-thiogalactopyranoside (IPTG).
6. Ni-NTA Agarose (Qiagen).
7. Affi-Gel 15 (Bio-Rad Laboratories).

8. Adenosine 5′-triphosphate, disodium salt.
9. [γ-^{32}P]ATP (Perkin Elmer).
10. Phosphoric acid.
11. Kinase-Glo Luminescent Kinase assay kit (Promega): Reconstitute fresh following the manufacturer's protocol prior to experiment.
12. Peptide and Soxtide (Sox-modified peptide) (peptide sequence: RKKYKFNEDTERRRFL) (*see* **Note 1**).

2.2 Buffers and Solutions

All stock solutions should be filtered with 0.2 μm cellulose nitrate membrane, and molecular biology-grade water was used in all solutions.

1. Lysis buffer: 20 mM Tris–HCl (pH 8.0), 0.5 M NaCl, 5 mM imidazole, 5 mM $MgCl_2$, 1 % (v/v) Triton X-100, 0.03 % (v/v) Brij 35, 0.1 % (v/v) β-mercaptoethanol, 1 mM benzamidine hydrochloride hydrate, 0.1 mM TPCK (Tosyl phenylalanyl chloromethyl ketone), 0.1 mM PMSF (phenylmethanesulfonyl fluoride), and 7 μM lysozyme.
2. Ni-NTA-column wash buffer: 20 mM Tris–HCl (pH 8.0), 0.25 M NaCl, 40 mM imidazole, 0.03 % (v/v) Brij 35, 0.1 % (v/v) β-mercaptoethanol, 1 mM benzamidine hydrochloride hydrate, 0.1 mM TPCK, and 0.1 mM PMSF.
3. Ni-NTA-column elution buffer: 20 mM Tris–HCl (pH 8.0), 0.25 M NaCl, 250 mM imidazole, 0.03 % (v/v) Brij 35, 0.1 % (v/v) β-mercaptoethanol, 1 mM benzamidine hydrochloride hydrate, 0.1 mM TPCK, and 0.1 mM PMSF.
4. TEV protease cleavage buffer and gel filtration chromatography buffer: 20 mM Tris–HCl (pH 8.0), 0.15 M NaCl, 5 mM $MgCl_2$, and 0.1 % (v/v) β-mercaptoethanol.
5. CaM-column coupling buffer: 50 mM HEPES (pH 7.5), 50 mM NaCl, 1 mM $CaCl_2$, and 0.1 % (v/v) β-mercaptoethanol.
6. CaM-column elution buffer: 50 mM HEPES (pH 7.5), 50 mM NaCl, 7.5 mM EGTA (ethylene glycol tetraacetic acid), and 0.1 % (v/v) β-mercaptoethanol.
7. Storage buffer: 25 mM HEPES (pH 7.5), 2 mM dithiothreitol (DTT), 50 mM KCl, 0.1 mM ethylenediaminetetraacetic acid (EDTA), 0.1 mM EGTA, and 10 % (v/v) glycerol.
8. Assay buffer: 25 mM HEPES (pH 7.5), 50 mM KCl, 0.1 mM EDTA, 0.1 mM, EGTA, 2 mM DTT, 150 μM $CaCl_2$, 1 μM CaM, 40 μg/mL BSA, and 10 mM $MgCl_2$.

2.3 Consumables and Equipment

1. ÄKTA FPLC™ System (Amersham Biosciences).
2. HiPrep™ 26/60 Sephacryl™ S-200 HR gel filtration column (GE Healthcare).
3. Cary 50 UV–Vis spectrophotometer.

4. Sonicator.
5. Ultracentrifuge (Beckman Coulter).
6. Microwell plates: 384-well black polystyrene plates (Nunc, fluorescence assay), 384-well white polystyrene plates (Nunc, luminescence assay), 96-well clear polystyrene plates (Corning, radioactive assay), p81 phosphocellulose plates (Whatman, radioactive assay scintillation count), 384-well polypropylene plates (Greiner) (*see* **Note 2**).
7. JANUS automated liquid handling workstation (Perkin Elmer).
8. EnVision 2103 Multilabel Reader (Perkin Elmer).
9. MicrofloSelect Bulk liquid dispenser (Biotek).
10. MicroBeta®TriLux scintillation counter (Perkin Elmer).
11. PlateLoc Plate sealer (Agilent, Santa Clara CA) loaded with peelable aluminum seal (Agilent).
12. High-speed centrifuge with swing-bucket rotors to accommodate standard microplates up to four plates per rotor (Eppendorf, #5810R).
13. Desiccator cabinet.
14. Bar-code printer (Zebra, Z4M Plus, *see* **Note 3**).
15. Twelve-channel pipette (Rainin).
16. Compound library (*see* **Note 4**): NIH clinical collection (446 compounds, Evotec), Spectrum collection (2000 compounds, Microsource Discovery System, Inc), Kinase set (11,250 compounds, Chembridge), Fragment Set (4000 compounds, Chembridge), Maybridge Hitfinder v9 (14,400 compounds, ThermoFisher), and Kinase Focused library (650 compounds, TxSACT, *see* **Note 5**).

3 Methods

3.1 eEF-2K Purification

1. Amplify human eEF-2K cDNA from the bacterial expression vector pGEX-2T that contains the gene encoding eEF-2K (GenBank accession number NM_013302) by using PCR with specific primers that bind to the 5′ and 3′ regions of the gene. The primers incorporate *Eco*RV (5′) and *Xho*I (3′) restriction sites to allow subcloning into the pET32a bacterial overexpression plasmid and to yield the *p32eEF-2K* plasmid.
2. To facilitate cleavage of the Trx-His$_6$-tag from recombinant eEF-2K, replace the sequence coding for the enterokinase (EK) cleavage site (DDDDK) in *p32eEF-2K* by one coding for the tobacco etch virus (TEV) protease cleavage recognition sequence (ENLYFQGDI) to generate *p32TeEF-2K* (*see* **Note 6**).

3. Sequence the recombinant human eEF-2K gene to verify the correct DNA sequence for the gene.
4. Grow *Escherichia coli* Rosetta-gami™2(DE3) cells transformed with *p32TeEF-2K* expression vector at 37 °C in 6 L of *E. coli* culture media (*see* **Note 7**).
5. Induce protein expression in the early logarithmic phase of cell growth (OD_{600} of 0.6–0.8) by the addition of 0.2 mM IPTG. Allow protein expression for 16 h at 22 °C and harvest the cells by centrifugation.
6. Resuspend bacterial pellet in 100 mL of lysis buffer. Lyse cells on ice by sonicating three times for 5 min (5-s pulses; 50:50 cycle) with an interval of 5 min. Perform all subsequent purification steps at 4 °C.
7. Clear lysate by centrifugation at 27,000 × *g* for 30 min.
8. After transferring to a clean tube, gently agitate supernatant with 5 mL of Ni-NTA beads for 1 h.
9. Load lysate onto a 100 mL chromatography column and wash beads with 10 column volumes of wash buffer.
10. Elute His-tagged protein with 30 mL of elution buffer.
11. Cleave the Trx-His_6-tag with 1.5 % TEV protease (w/w) for 4 h while simultaneously dialyzing the sample against TEV protease cleavage buffer (*see* **Note 8**).
12. Separate kinase from cleaved tag with a CaM-agarose affinity column (*see* **Note 9**). Add $CaCl_2$ to sample to a final concentration of 1 mM and gently agitate with 5 mL of CaM-agarose beads for 1 h.
13. Load sample onto a 50 mL chromatography column and wash beads with 10 column volumes of CaM-column coupling buffer.
14. Elute tagless kinase with 15 mL of CaM-column elution buffer.
15. Concentrate eluate to a volume of 5 mL and apply to a HiPrep™ 26/60 Sephacryl™ S-200 HR gel filtration column. Perform chromatography over one column volume (320 mL) at a flow rate of 1 mL/min. Identify highly purified monomeric fractions using sodium dodecyl sulfate-polyacrylamide gel electrophoresis (SDS-PAGE) (*see* **Note 10**).
16. Pool and dialyze against storage buffer.
17. Determine concentration of eEF-2K by measuring absorbance at 280 nm and using an extinction coefficient (A_{280}) of 97150 cm^{-1} M^{-1} (which is calculated from the primary amino acid sequence).
18. Aliquot, flash freeze in liquid nitrogen, and store at −80 °C until ready to use.

3.2 Preparation of Compound Dilution

1. Use a bulk dispenser to fill dilution plates with 40 μL of water (*see* **Note 11**).
2. Use Janus liquid handler with a 384-channel pipetting head to transfer 1 μL of compound from a library plate to a dilution plate (*see* **Note 11**).

3.3 eEF-2K Kinase Assays

3.3.1 Fluorescence-Based High-Throughput Screen

Unless otherwise mentioned, plates were spun down at 600 rpm for 1 min after every liquid transfer. All assays were conducted at room temperature. Fluorescence- and luminescence-based assays were employed for primary and cherry pick screens, while a radioactive assay was utilized as an orthogonal screen and dose-dependency profile. A primary screen was conducted in singleton, while cherry pick and orthogonal screens were conducted in duplicate. The overall screen flow is profiled in Fig. 2. Assay optimization and miniaturization were detailed in a previous report [23].

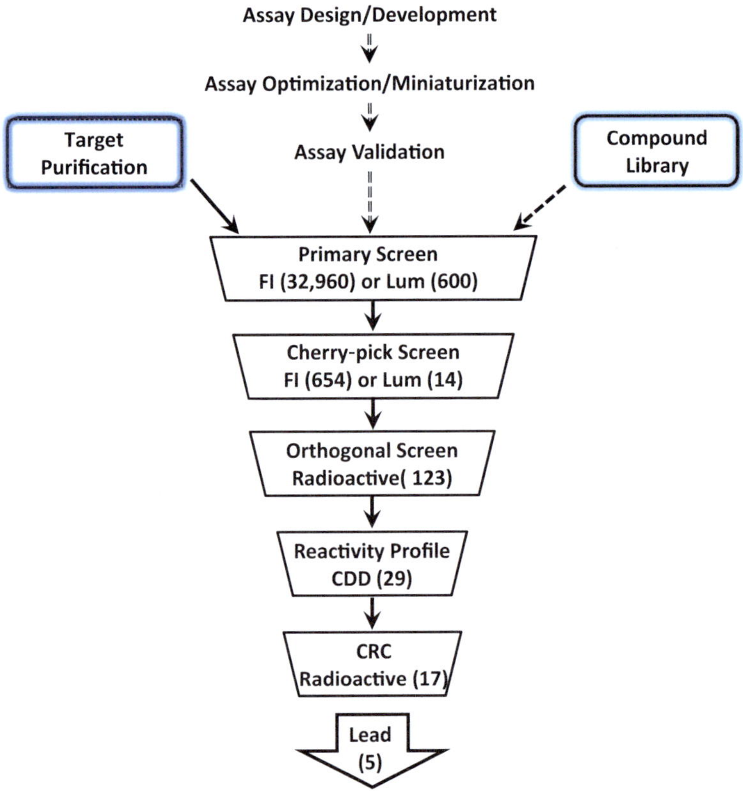

Fig. 2 Profile of inhibitor screening: Numbers in parentheses represent numbers of compounds tested on corresponding procedures. Only solid arrows are described in this chapter. In our particular screening procedure, five compounds were identified as "lead" for in vivo study and further optimization. FI and Lum represent fluorescence- and luminescence-based assays

1. Prepare assay mixture by diluting eEF-2K, Soxtide, and Brij-35 in assay buffer to a concentration of 14 nM, 14 μM, and 0.042 % (v/v), respectively. Final concentrations of eEF-2K, Soxtide, and Brij-35 will be 10 nM, 10 μM, and 0.03 % (v/v), respectively (*see* **Note 12**).
2. Make 500 μM ATP solution in assay buffer.
3. Use a Janus liquid handler to transfer 7 μL assay mixture to assay plates. Fill columns 1–22 with assay mixture while filling columns 23 and 24 with assay mixture excluding enzyme (*see* **Note 13** and Fig. 3).
4. Use a Janus liquid handler to transfer 1 μL aliquots of compound dilution to assay plates and incubate for 30 min.
5. Dispense 2 μL of ATP using a bulk dispenser to initiate enzymatic reaction and monitor relative fluorescence for 5 min with 1-min intervals using a plate reader equipped with excitation and emission filters at 355 nm and 460 nm, respectively.
6. Calculate initial velocities for each compound tested using a formula: $F(t) = -vt + F_0$, where t, time; $F(t)$, fluorescence intensity at time t; F_0, fluorescence intensity at $t = 0$; and v, initial velocity, and use for further analysis (% inhibition and z' calculation, *see* **Note 14**).

3.3.2 Luminescence-Based High-Throughput Screen

1. Prepare assay mixture by diluting eEF-2K and peptide in assay buffer to concentrations of 10 nM and 60 μM, respectively. Final concentrations are 5 nM and 30 μM, respectively.
2. Prepare assay plates by transferring 5 μL assay mixture to assay plates: *see* Subheading 3.3.1, **step 3**.

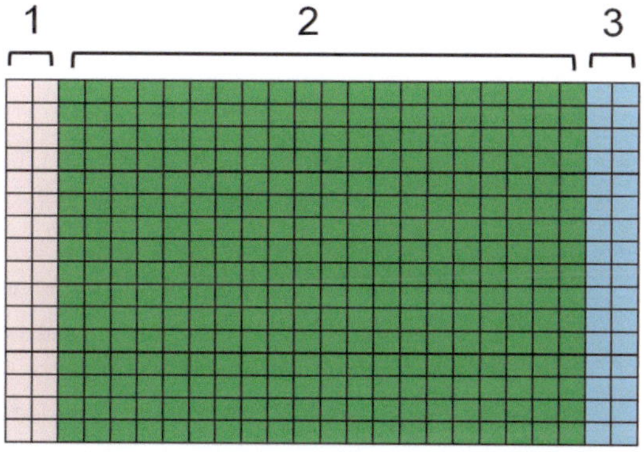

ID	Categories	# Wells	Components	
1	Negative Control	32	Enzyme + Peptide	DMSO
3	Positive Control	32	Peptide	DMSO
2	Samples	320	Enzyme + Peptide	DMSO + Compound

Fig. 3 Typical assay plate format in a 384-well plate for high-throughput screening

3. Transfer compound dilution and incubate: *see* Subheading 3.3.1, **step 4**.
4. Transfer 4 μL of 2.5 μM ATP solution (final concentration of 1 μM) to assay plates using a bulk dispenser to initiate reaction, and incubate for 80 min.
5. Transfer 10 μL of the Kinase-Glo Luminescent Kinase assay kit using Janus, incubate for 10 min, and measure the luminescence counts on a plate reader.
6. Calculate % inhibition and z' values (*see* **Note 15**).

3.3.3 Cherry-Pick Screen

1. Fill the dilution plate with 40 μL of water. Cherry-pick 1 μL of compounds which were identified as hits from primary screen and transfer to the corresponding wells on the dilution plate using a Janus liquid handler (*see* **Note 16**).
2. Rescreen diluted compounds using either a fluorescence- or a luminescence-based screen.
3. Calculate reproducibility using a formula: Reproducibility = 100*(% Inh_{ch}/% Inh_{pr}), where Inh_{ch} = % inhibition from cherry-pick screen and Inh_{pr} = % inhibition from primary screen.

3.3.4 Radioactive Assay

A radioactivity assay was employed twice: first to eliminate false hits due to compounds' interference with analytical output or with assay components (orthogonal screen, Fig. 2), and second to examine compound potency (i.e., IC_{50} or compounds-response curve, referred to as "CRC" in Fig. 2).

1. Prepare assay mixture by diluting enzyme and peptide (2 nM and 50 μM at final concentrations, respectively) in assay buffer.
2. Prepare compounds in dilution plates: *see* Subheading 3.3.3, **step 1** (*see* **Note 17**).
3. Prepare assay plates by transferring 10 μL of diluted compounds to columns 2–11 and 10 μL DMSO diluted in water to columns 1 and 12 as controls, using a multichannel pipette. Dispense 70 μL of the assay mixture to columns 1–11 and positive control assay mixture to column 12 (no enzyme). Compounds were screened at 50 μM for all libraries except for fragment, which was screened at 100 μM.
4. Incubate assay plates at room temperature for 30 min.
5. Add 20 μL of [γ-^{32}P]ATP to assay plates to achieve the final concentration of 100 μM.
6. After 10 min, stop the reaction by adding 100 μL of 100 mM phosphoric acid.

7. Transfer 150 μL of assay mixtures to the p81 phosphocellulose plates using a 12-channel pipette. Wash the plate ten times with 50 mM phosphoric acid to get rid of excess ATP and dry with acetone.
8. Read the scintillation counts using a scintillation counter in the presence of 20 μL scintillation fluid per well and calculate % inhibition (similar formula in *see* **Note 14**).
9. Compounds demonstrating higher than 30 % activity were analyzed for reactivity profiles to exclude compounds containing toxic or reactive features (*see* **Note 18**). Among them, compounds satisfying the Lipinski's rule of 5 were purchased and re-assayed using the radioactive assay at several concentrations of the inhibitors.
10. Calculate IC_{50} (inhibitor concentration required to achieve 50 % inhibition) and Hill coefficient using a 4-parameter logistic nonlinear regression from Kaleidagraph 4.0 (*see* **Note 19**).

4 Notes

1. Details of peptide synthesis and purification were described elsewhere [23].
2. Choice of plate material, size, and color is very critical in maximizing signal and reducing nonspecific adhesion to the plastics. For in vitro assays, polystyrene is recommended because biological reagents are less prone to adhere to it than to polypropylene. Depending on the assay volume, regular well (25–90 μL in 384-well format) or low-volume well (<20 μL in 384-well format) can be chosen. Black plates are better for fluorescence, while white is better for luminescence. Dilution plates are used to dilute compounds prior to addition to the assay plate. Because compounds are generally stored at 10 mM concentration in 100 % DMSO and the lowest volume that conventional liquid handlers can handle is ~ 0.5 μL, compound needs to be diluted twice to meet desired concentration. For the first dilution, any type of polypropylene plate can be used and usually compounds are diluted 100–200-fold in water.
3. Bar code is essential to track compounds in the high-throughput process. It was found to be practical that dilution and assay plates were all labeled with the same bar codes corresponding to the testing library plates.
4. All compounds were formatted in 384-well plates. All library plates contained pure DMSO (without compounds) in columns 1, 2, 23, and 24 to be used as controls. *Caution for library plate handling*: Library plates were thawed inside a desiccator cabinet overnight and spun down at 600 rpm for 1 min

at room temperature prior to use. After use, plates were immediately purged with argon gas for 10 s, heat sealed, and stored at −40 °C. All library plates were bar-coded and are retired after ~10 freeze-thaw cycles.

5. This unique collection was custom-assembled by Texas Screening Alliance for Cancer Therapeutics (TxSACT), comprised of more than 600 small molecules with known activities against around 100 kinases. This collection represents a useful panel of compounds for developing leads against novel ATP-dependent proteins based on their off-target activities.
6. Design oligonucleotides to match a *Kpn*I-*Eco*RV-digested plasmid and include a sequence coding for the TEV protease cleavage recognition site between the restriction sites. Mix both oligonucleotides in equimolar amounts, heat to 95 °C for 5 min, and cool to allow annealing in order to obtain a double-stranded DNA fragment with blunt and sticky ends. Ligate this fragment into the *p32eEF-2K* construct digested with *Kpn*I and *Eco*RV.
7. Out of a total of 725 residues, eEF-2K contains 63 rare codons (8.7 %). To potentially alleviate codon bias during expression, use the Rosetta-gami 2(DE3) strain which carries the pRARE2 plasmid (supplies tRNAs for seven rare codons).
8. Transform the construct encoding tobacco etch virus protease (pRK793 expression vector) into Rosetta-gami™ 2(DE3) cells and induce protein expression with 0.5 mM IPTG at 28 °C for 4 h before harvesting. Purify the protein according to protocols published earlier [24, 25].
9. Transform the construct encoding CaM (pET-23 expression vector) into BL21(DE3) cells (Novagen) and induce protein expression with 0.5 mM IPTG at 30 °C for 5 h. Purify CaM as previously described [26] with minor modifications [27]. Dialyze the purified CaM against 100 mM HEPES (pH 7.5) and concentrate to 1.5 mM. Generate CaM-agarose beads by coupling CaM to Affi-Gel 15-activated affinity media as per the manufacturer's protocol. Gently couple Affi-Gel 15 (5 mL) with 7.5 mL of CaM on a shaker at 4 °C for 4 h. Wash beads with 10 column volumes of CaM-column coupling buffer and store at 4 °C in the same buffer containing 0.2 % (w/v) sodium azide.
10. Perform this additional purification step with the aim of separating out monomeric eEF-2K from eEF-2K aggregates. When separated on a Sephacryl gel filtration column, monomeric eEF-2K is most likely eluted in the second peak (110–130 mL) since it possesses a fourfold higher activity than the enzyme eluted in the first peak [28].

11. Depending on dilution ratio and considering liquid handlers' dispensing limitation, vary the volumes of water and dispense compounds accordingly. Unless otherwise specified, all the compounds were screened at 25 μM except for fragment library (100 μM). Prepare compound dilution fresh before use.

12. Make up a bulk assay mix sufficient for the total number of assay plates per screening experiment. Bulk preparation is beneficial in improved reproducibility between assay plates. Extra volume for reagent reservoir should be considered (e.g., 12 column reagent reservoir requires ~0.6 mL of dead volume per column.).

13. Each of the two columns at both ends on an assay plate is reserved for negative (0 % inhibition, columns 1 and 2) and positive (100 % inhibition, columns 23 and 24) controls, respectively. Locating both controls per assay plate is more worthwhile because the intra-plate variation is more critical than inter-plate variation in determining compounds' activities. Pairs of columns 1/24 and columns 2/23 can be used as negative and positive controls, respectively, if edge effect is a concern. A typical 384-well assay plate format for high-throughput screening is illustrated in Fig. 3.

14. Percent inhibition for each compound tested was calculated using a formula: % Inhibition = 100 * $(v_{\text{Positive control}} - v_{\text{Compounds}})/(v_{\text{Negative control}} - v_{\text{Positive control}})$. Data obtained from both controls are used to calculate z' [29] per assay plate and to determine assay quality. Calculated z' values from fluorescence- and luminescence-based assays were in the range of 0.75 ± 0.1 and 0.80–0.85, respectively. Kinetic measurement is often informative for the compounds with higher inherent fluorescence. In addition, the signal-to-background ratio (S/B) and the signal-to-noise ratio (S/N) were almost 70:1 and 21:1, respectively, when measured as an initial velocity, while relative fluorescence intensity after 5 min was only around 1.5-fold higher than the background. In this regard, we consider kinetic measurement as more beneficial than absolute intensity assessment.

15. *See* **Note 14**. Use the same formula for % inhibition calculation substituting initial velocity by luminescence count.

16. We used CDD's program (Collaborative Drug Discovery, San Francisco, CA) to rank compounds' activity based on % inhibition and selected compounds showing >50 % inhibition as primary hits for further confirmation screening. Compound ranking and hit criteria can be done in various ways as mentioned elsewhere [30]. The cherry-pick is necessary to quickly remove false hits due to errors from plastics or assay limitations.

17. Compounds showing reproducible activities (>30 %) in a cherry-pick screen were tested in an orthogonal screen.
18. Compounds containing highly reactive functional groups that are likely or known to interact with proteins in a nonspecific manner were eliminated from further analysis.
19. It is useful to examine CRC using either (or both) fluorescence/luminescence-based screens to validate IC_{50}values obtained from radioactive assays.

Acknowledgements

This research was supported in part by the grants from the Welch Foundation (F-1390), CPRIT (RP110532-P1), and NIH (GM059802 and CA167505).

References

1. Wu H, Yang JM, Jin S et al (2006) Elongation factor-2 kinase regulates autophagy in human glioblastoma cells. Cancer Res 66:3015–3023
2. Hait WN, Wu H, Jin S et al (2006) Elongation factor-2 kinase: its role in protein synthesis and autophagy. Autophagy 2:294–296
3. Calberg U, Nilsson A, Skog S et al (1991) Increased activity of the eEF-2 specific, Ca^{2+} and calmodulin dependent protein kinase III during the S-phase in Ehrlich ascites cells. Biochem Biophys Res Commun 180:1372–1376
4. Parmer TG, Ward MD, Yurkow EJ et al (1999) Activity and regulation by growth factors of calmodulin-dependent protein kinase III (elongation factor 2-kinase) in human breast cancer. Br J Cancer 79:59–64
5. Bagaglio DM, Cheng EH, Gorelick FS et al (1993) Phosphorylation of elongation factor 2 in normal and malignant rat glial cells. Cancer Res 53:2260–2264
6. Hait WN, Ward MD, Trakht IN et al (1996) Elongation factor-2 kinase: immunological evidence for the existence of tissue-specific isoforms. FEBS Lett 397:55–60
7. Parmer TG, Ward MD, Hait WN (1997) Effects of rottlerin, an inhibitor of calmodulin-dependent protein kinase III, on cellular proliferation, viability, and cell cycle distribution in malignant glioma cells. Cell Growth Differ 8:327–334
8. Cheng Y, Li H, Ren X et al (2010) Cytoprotective effect of the elongation factor-2 kinase-mediated autophagy in breast cancer cells subjected to growth factor inhibition. PLoS One 5, e9715
9. Wu H, Zhu H, Liu DX et al (2009) Silencing of elongation factor-2 kinase potentiates the effect of 2-deoxy-D-glucose against human glioma cells through blunting of autophagy. Cancer Res 69:2453–2460
10. Tekedereli I, Alpay SN, Tavares CD et al (2012) Targeted silencing of elongation factor 2 kinase suppresses growth and sensitizes tumors to doxorubicin in an orthotopic model of breast cancer. PLoS One 7, e41171
11. Autry AE, Adachi M, Nosyreva E et al (2011) NMDA receptor blockade at rest triggers rapid behavioural antidepressant responses. Nature 475:91–95
12. Leprivier G, Remke M, Rotblat B et al (2013) The eEF2 Kinase Confers Resistance to Nutrient Deprivation by Blocking Translation Elongation. Cell 153:1064–1079
13. Arora S, Yang JM, Kinzy TG et al (2003) Identification and characterization of an inhibitor of eukaryotic elongation factor 2 kinase against human cancer cell lines. Cancer Res 63:6894–6899
14. Devkota AK, Tavares CD, Warthaka M et al (2012) Investigating the kinetic mechanism of inhibition of elongation factor 2 kinase by NH125: evidence of a common in vitro artifact. Biochemistry 51:2100–2112
15. Chen Z, Gopalakrishnan SM, Bui MH et al (2011) 1-Benzyl-3-cetyl-2-methylimidazolium

iodide (NH125) induces phosphorylation of eukaryotic elongation factor-2 (eEF2): a cautionary note on the anticancer mechanism of an eEF2 kinase inhibitor. J Biol Chem 286:43951–43958

16. Parker MW, Lo BM, Federici G (1990) Crystallization of glutathione S-transferase from human placenta. J Mol Biol 213:221–222
17. https://www.gelifesciences.com/gehcls_images/GELS/Related%20Content/Files/1336168762999/litdoc18114275_20130524112435.pdf
18. von Ahsen O, Bomer U (2005) High-throughput screening for kinase inhibitors. Chembiochem 6:481–490
19. Shults MD, Imperiali B (2003) Versatile fluorescence probes of protein kinase activity. J Am Chem Soc 125:14248–14249
20. Shults MD, Janes KA, Lauffenburger DA et al (2005) A multiplexed homogeneous fluorescence-based assay for protein kinase activity in cell lysates. Nat Methods 2:277–283
21. http://tools.lifetechnologies.com/content/sfs/manuals/omnia_kinase_assay_man.pdf
22. Tavares CD, O'Brien JP, Abramczyk O et al (2012) Calcium/calmodulin stimulates the autophosphorylation of elongation factor 2 kinase on Thr-348 and Ser-500 to regulate its activity and calcium dependence. Biochemistry 51:2232–2245
23. Devkota AK, Warthaka M, Edupuganti R et al (2014) High-Throughput Screens for eEF-2 Kinase. J Biomol Screen 19:445–452
24. Kapust RB, Tozser J, Fox JD et al (2001) Tobacco etch virus protease: mechanism of autolysis and rational design of stable mutants with wild-type catalytic proficiency. Protein Eng 14:993–1000
25. Kristelly R, Earnest BT, Krishnamoorthy L et al (2003) Preliminary structure analysis of the DH/PH domains of leukemia-associated RhoGEF. Acta Crystallographica Sect D 59:1859–1862
26. Gaertner TR, Kolodziej SJ, Wang D et al (2004) Comparative Analyses of the Three-dimensional Structures and Enzymatic Properties of α, β, γ, and δ Isoforms of Ca^{2+}-Calmodulin-dependent Protein Kinase II. J Biol Chem 279:12484–12494
27. Forest A, Swulius MT, Tse JKY et al (2008) Role of the N- and C-Lobes of Calmodulin in the Activation of Ca^{2+}/Calmodulin-Dependent Protein Kinase II. Biochemistry 47:10587–10599
28. Abramczyk O, Tavares CDJ, Devkota AK et al (2011) Purification and characterization of tagless recombinant human elongation factor 2 kinase (eEF-2K) expressed in Escherichia coli. Protein Expr Purif 79:237–244
29. Zhang JH, Chung TD, Oldenburg KR (1999) A Simple Statistical Parameter for Use in Evaluation and Validation of High Throughput Screening Assays. J Biomol Screen 4:67–73
30. Taylor JA, Burton NP, Maxwell A (2012) High-throughput microtitre plate-based assay for DNA topoisomerases. Methods Mol Biol 815:229–239

Chapter 3

Recombinant Kinase Production and Fragment Screening by NMR Spectroscopy

Byeonggu Han and Hee-Chul Ahn

Abstract

During the past decade fragment-based drug discovery (FBDD) has rapidly evolved and several drugs or drug candidates developed by FBDD approach are clinically in use or in clinical trials. For example, vemurafenib, a V600E mutated BRAF inhibitor, was developed by utilizing FBDD approach and approved by FDA in 2011. In FBDD, screening of fragments is the starting step for identification of hits and lead generation. Fragment screening usually relies on biophysical techniques by which the protein-bound small molecules can be detected. NMR spectroscopy has been extensively used to study the molecular interaction between the protein and the ligand, and has many advantages in fragment screening over other biophysical techniques. This chapter describes the practical aspects of fragment screening by saturation transfer difference NMR.

Key words Fragment, Fragment-based drug discovery, Kinase, Optimum solubility screening, NMR screening, Saturation transfer difference, Surface plasmon resonance, X-ray crystallography

1 Introduction

Kinases are implicated in varieties of diseases, such as, cancer, immunological, neurological, and metabolic diseases, and thus are one of the most important therapeutic targets in drug discovery. More than 20 drugs targeting kinases have been approved and more than 100 kinase inhibitors are under the clinical trials. Most of the initial hits against kinases were developed with a conventional drug discovery campaign including high-throughput screening which is typically based on biochemical and cellular assays. Structure-guided approaches have been combined and provided a powerful insight into the selectivity enhancement and lead optimization. Overviews and perspectives in discovering new kinase inhibitors were recently reviewed [1, 2].

Hicham Zegzouti and Said A. Goueli (eds.), *Kinase Screening and Profiling: Methods and Protocols*, Methods in Molecular Biology, vol. 1360, DOI 10.1007/978-1-4939-3073-9_3, © Springer Science+Business Media New York 2016

As an alternative approach to high-throughput screening, the fragment screening has become a valuable tool for the generation of novel chemical leads. Since the first report appeared in 1996 describing structure activity relationship (SAR) by NMR [3], FBDD has been widely utilized by pharmaceutical companies as well as academic research institutes to discover small molecular inhibitors against varieties of therapeutic targets [4]. SAR by NMR includes identification of small organic compounds which bind to proximal sites on target proteins and followed by the optimization and (or) tethering of individual compounds to produce high-affinity ligands [3]. Biophysical technique by which binding of the compound to the target protein is monitored can be employed for fragment screening. Not only NMR spectroscopy but also isothermal titration calorimetry (ITC), surface plasmon resonance (SPR), mass spectrometry, and X-ray crystallography are extensively utilized in the screening.

For example, Plexxikon discovered vemurafenib (Zelboraf™) for the treatment of late-stage melanoma targeting the mutated form of the kinase B-RAF (V600E) by a combination of scaffold- and structure-based screening [5, 6]. Astex applied FBDD strategy to the molecular chaperone HSP90 of which client proteins are the therapeutic targets of cancer, resulting in compounds that are currently under clinical trials [7–9]. Application of FBDD strategy has not been restricted to development of anticancer drugs. Fragment screening with 2D NMR followed by X-ray structure determination was conducted on the aspartic protease β-secretase (BACE) for the treatment of Alzheimer's disease, which resulted in the discovery of MK-8931 [10, 11]. Likewise, FBDD have been growing rapidly and was widely used by pharmaceutical and academic institutions, and increasing number of compounds originating from FBDD approaches are under development or in clinical trials [1, 2, 12].

FBDD starts with the screening of chemical library, called fragment screening, and at this point the detection of direct binding between the fragment and the target protein is the key step to identification of hits. For this purpose, the preparation of large amount of the target protein is prerequisite and appropriate detection methods of the protein-ligand binding are needed. Of those biophysical techniques to identify the protein-bound chemicals, NMR spectroscopy has some advantages especially in the characterization of the reversible binding of weak affinity small molecules to the target proteins. Since characteristics of NMR signals of compounds from free- and protein-bound forms are quite different, it is easy and straightforward to identify certain chemical(s) which can bind to the target protein from a pool of chemicals. Many NMR techniques have been evolved so far and provide essential information in the different stages of drug discovery process, and were summarized in many review articles [13, 14].

Saturation-transfer difference (STD) NMR has been widely used to detect the binding of a small ligand to its receptor and to identify the binding epitope of a ligand when it is bound to its receptor protein [14–17]. The selective saturation of protons in the macromolecular receptor by irradiating the ligand-free spectral region evolves the spin diffusion across the entire receptor. At the same time the spin diffusion also spread onto the ligand if the small molecule is bound to the receptor, which result in the attenuation of the intensity of the ligand signal. Subtracting on-resonance spectrum from off-resonance spectrum (without saturation) yields the difference spectrum where only resonances of the bound ligands can remain (Fig. 1). This chapter focuses on the early stages

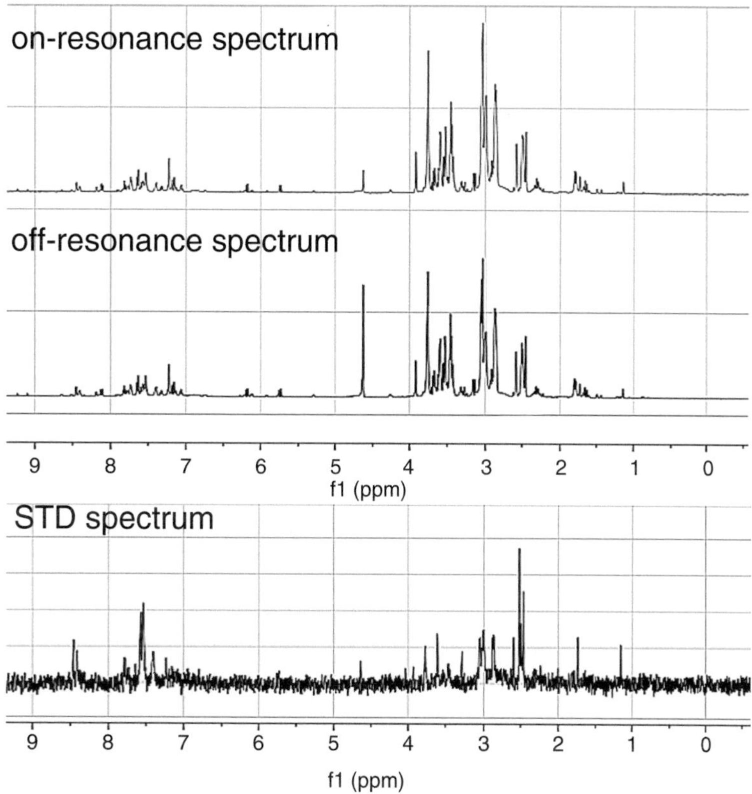

Fig. 1 STD NMR spectrum of JNK3 catalytic domain with 8-compound mixture. Subtracting on-resonance spectrum (*top*) from off-resonance spectrum (*middle*) yielded STD spectrum (*bottom*). The signal intensities of JNK3-bound compounds were attenuated in on-resonance spectrum whereas the intensities of non-binder signals were not affected by saturation of the receptor (see text and Subheading 4). The STD spectrum, the resultant of subtracting on-resonance spectrum from off-resonance spectrum, clearly shows the resonances from JNK3-bound compounds. The resonances at 4.6 ppm on each spectrum are residual H_2O signals. Vertical scales of on-resonance and off-resonance spectra are equivalent; however, that of STD spectrum is arbitrary

in FBDD targeting kinases including the methods of recombinant protein production in *E. coli* and insect cells, and fragment screening by STD NMR.

2 Materials

2.1 Cloning and Expression of Recombinant Kinase

1. pET-15b vector (Novagen).
2. *E. coli* strains such as BL21(DE3), Rossetta 2 (DE3), C41 (DE3), pLysS (DE3), and Arctic Express (DE3) cells.
3. IPTG: 0.1 mM isopropyl-β-D-thiogalactopyranoside.
4. pAcHLT-A Baculovirus transfer vector (PharMingen).
5. BaculoGold DNA (PharMingen).

2.2 Buffers

1. Lysis buffer: 20 mM HEPES (pH 7.4), 150 mM NaCl, 10 % glycerol, 1 mM TCEP, and EDTA-free protease inhibitor cocktail (Roche).
2. Binding buffer: 20 mM HEPES (pH 7.4), 150 mM NaCl, 10 % glycerol, 1 mM TCEP, and 10 mM imidazole.
3. Washing buffer: 20 mM HEPES (pH 7.4), 150 mM NaCl, 10 % glycerol, 1 mM TCEP, and 20 mM imidazole.
4. Elution buffer: 20 mM HEPES (pH 7.4), 150 mM NaCl, 10 % glycerol, 1 mM TCEP, and 500 mM imidazole.
5. SEC buffer: 20 mM HEPES (pH 7.4), 150 mM NaCl, 10 mM $MgCl_2$, 1 mM EDTA, and 2 mM TCEP.
6. Storage buffer: 20 mM selected buffer, 100 mM NaCl, 10 mM $MgCl_2$, 1 mM EDTA, and 2 mM TCEP
7. Solubility test buffers in Table 1.
8. NMR screening buffer: 10 mM $_{d11}$-Tris-Cl buffer, pH 7.5 in 100 % D_2O.
9. JNK3 Buffer: 20 mM sodium phosphate buffer, pH 7.2, 100 mM NaCl, and 2 mM DTT.

2.3 Equipment and Consumables

1. French Press.
2. Dounce homogenizer.
3. Dynamic light scattering (DLS).
4. HiLoad 16/600 Superdex 75 prep-grade column (GE Healthcare).
5. Amicon Ultra (Millipore).
6. NMR spectrometer.

2.4 Software

NMR data processing software such as MnovaNMR (Mestrelab Research) or ACD/NMR Processor (Advanced Chemistry Development Inc.)

Table 1
Solubility test buffers

Number	Buffer (100 mM)	pH
1	Citric acid	4.0
2	Sodium acetate	4.5
3	Sodium/potassium phosphate	5.0
4	Sodium citrate	5.5
5	Sodium/potassium phosphate	6.0
6	Bis-tris	6.0
7	MES	6.2
8	ADA	6.5
9	Cacodylate	6.5
10	MOPS	7.0
11	Sodium/potassium phosphate	7.0
12	HEPES	7.0
13	HEPES	7.5
14	Tris	7.5
15	Imidazole	8.0
16	Tris	8.5

3 Methods

3.1 Recombinant Kinase Production in E. coli

Protein expression in *E. coli* has been the first choice among many methods of the recombinant protein production since the cloning is rapid, the expression cost is low, and there are plenty of commercially and academically available sources like expression kits, vectors, and host cell lines. Here we describe procedures to produce the catalytic domain of c-Jun N-terminal kinase (JNK) isoform 3 in *E. coli*.

1. A DNA fragment encoding the kinase domain of human JNK3 spanning from serine 40 to glutamic acid 402 was amplified by PCR and subcloned into the NcoI and BamH1 sites of pET-15b vector. The fidelity of the plasmid was confirmed by DNA sequencing.
2. The plasmid containing the catalytic domain of JNK3 was transformed into a competent *E. coli* strain of choice. Transformed cell was grown in LB media for 16 h at 18 °C with IPTG.

3. Cells were harvested by centrifugation (3220 × *g*) at 4 °C for 30 min.
4. The collected cells were re-suspended in Lysis buffer and stored at −80 °C.
5. Aliquot of the harvested cell was subjected to SDS-PAGE and immunoblotting with anti-His antibody to measure the extent of protein expression.

3.2 Cloning and Expression Test with Baculovirus Expression System

Usually protein kinases and their domains are hardly expressed in prokaryotic cells. An alternative production of kinase is the expression of the protein in eukaryotic cells, such as Sf9 cell line.

1. The catalytic domain of kinase was subcloned into pAcHLT-A Baculovirus transfer vector and followed by DNA sequencing.
2. The plasmids containing the kinase (Baculovirus Transfer Vector) and BaculoGold DNA were co-transfected into insect cells (Sf9).
3. After 5 days, the supernatant of the co-transfection plates was collected. Transfected supernatants were treated to fresh Sf9 cells to produce high titer virus stocks. High titer viruses were used for target protein expression test by end-point dilution assay.
4. Cells were harvested and resuspended in lysis buffer and stored at −80 °C.

3.3 Purification of Recombinant Kinase

1. Re-suspended *E. coli* or Sf9 cells were thawed and lysed by French press or Dounce homogenizer on ice and centrifuged at 18,000 × *g* for 40 min at 4 °C.
2. The supernatant was diluted twofold with binding buffer I and applied to Ni-NTA column at 4 °C.
3. The column was washed with two column volumes of binding buffer I.
4. The column was washed with two column volumes of washing buffer.
5. The catalytic domain of JNK3 bound to Ni-NTA resin was eluted with five column volumes of elution buffer I.
6. Size-exclusion chromatography (SEC) was followed to achieve higher protein purity on a HiLoad 16/600 Superdex 75 prep-grade column pre-equilibrated with SEC buffer. The purified protein was concentrated by Amicon Ultra up to 10 mg/ml and stored at −80 °C.

3.4 Optimum Solubility Screening

Optimum solubility screening was originally suggested to optimize the buffer with high protein solubility and to obtain pure and conformationally homogenous protein samples, which is very critical in the preparation of protein samples for structural studies by X-ray crystallography [18]. The method includes selection of a better buffer than used during purification and the solubility of the protein is inspected by optical microscopy and DLS. We modified the original method for optimizing the storage buffer of kinase domain of JNK3 and for the use in fragment screening by STD NMR.

1. Prepare a set of 16 buffers (solubility test buffers, Table 1) with pH range from 4 to 8.5.
2. Dilute 5 μl of the concentrated protein solution into 45 μl of solubility test buffer.
3. Incubate each sample at room temperature for 1 h.
4. Centrifuge the sample with 15,000 × *g* at 4 °C for 10 min.
5. Monitor the total and soluble proteins by SDS-PAGE
6. Measure the monodispersity of the soluble proteins by DLS.
7. Select the protein samples with high solubility from SDS-PAGE and high monodispersity from DLS experiment.
8. Once a proper buffer was selected, modify the buffer ingredients and repeat the procedure 2–7 to optimize the final buffer. For JNK3 catalytic domain, sodium phosphate buffer, pH 7.0, was selected and the final JNK3 buffer was optimized to 20 mM sodium phosphate buffer, pH 7.2, 100 mM NaCl, and 2 mM DTT.

3.5 Fragment Library

Fragment library is the collection of small organic molecules, which covers diverse chemical spaces and druggable characteristics [19]. The criteria of selecting fragments are according to "Rule of Three (RO3)" (MW ≤ 300 Da, the number of H-bond donors ≤ 3, the number of H-bond acceptors ≤ 3, and cLogP ≤ 3). The number of rotatable bond, calculated total polar surface area (tPSA), and calculated water solubility (cLogSw) can be included in the selection criteria. There are many commercially available sources of fragment libraries such as Maybridge RO3 sets and ChemBridge 5000 RO3 fragments.

We have constructed our own fragment library, even though all of the compounds we constructed did not meet the RO3 criteria. The procedures were as follows.

1. Virtual screening of 15,000 compounds which have molecular weight less than 300 Da from Asinex, MayBridge, and ChemDive with the inhibitor bound JNK3 structures, PDB 3OY1 [20], 3DA6 [21], and 2EXC. The Glide program in the software Schrödinger suite was used.
2. Purchase of 492 compounds with the highest score from the Glide program.

3. Add $_{d6}$-DMSO to the compounds and store stock solutions at −20 °C. The final concentration of each stock solution was 80 mM (*see* **Note 1**).

3.6 Fragment Screening by STD NMR

1. Prepare 10 mg/ml JNK3 catalytic domain in JNK3 buffer. The final concentration of JNK3 catalytic domain was about 250 μM.
2. Dilute the fragment stock solution with NMR screening buffer. The final concentration of each compound was 320 μM and the volume of sample was 500 μl.
3. Measure the 1D ^{1}H NMR spectra of the 492 compounds individually; all NMR experiments were carried out at 25 °C using Agilent Unity 600 spectrometer equipped with z-axis gradient unit. 32 or 64 scans were enough to get a high quality spectra (*see* **Note 2**).
4. Add 1 μl of 8 fragment compound stocks into one Eppendorf tube and add 420 μl of NMR screening buffer; the concentration of each compound was 160 μM.
5. Measure the 1D ^{1}H NMR spectra of 8 compound mixtures; the number of samples is 62 at this point.
6. Add 10 μl of JNK3 solution to the NMR sample containing 8 compounds (the sample used in the procedure 5); the concentration of JNK3 catalytic domain in each NMR tube is about 5 μM.
7. Measure the 1D ^{1}H NMR spectra of the protein/compound mixtures. Since NMR samples contain about 1.1 M H_2O (50-fold dilution of JNK3 solution into the deuterated NMR screening buffer), water suppression during the NMR measurement is necessary to remove the residual H_2O signal. We employed Watergate water suppression during the experiments.
8. Measure STD NMR spectra with 62 NMR samples, each of which contains 5 μM proteins and 8 compounds (160 μM each). Before running the NMR experiments with the samples, it is recommended to check the efficiency of STD NMR with a known JNK3 binder, such as SP600125 [22] or the substrate ATP molecule (*see* **Notes 3** and **4**). Apply 512 or more scans to get the STD NMR spectrum.
9. Process NMR data with the software MnovaNMR (Mestrelab Research, http://Mestrelab.com) or ACD/NMR Processor (Advanced Chemistry Development Inc., http://www.acdlabs.com).
10. Compare the STD NMR spectra with the individual spectrum of fragments; Fig. 2 shows the identified fragments bound to JNK3 catalytic domain.

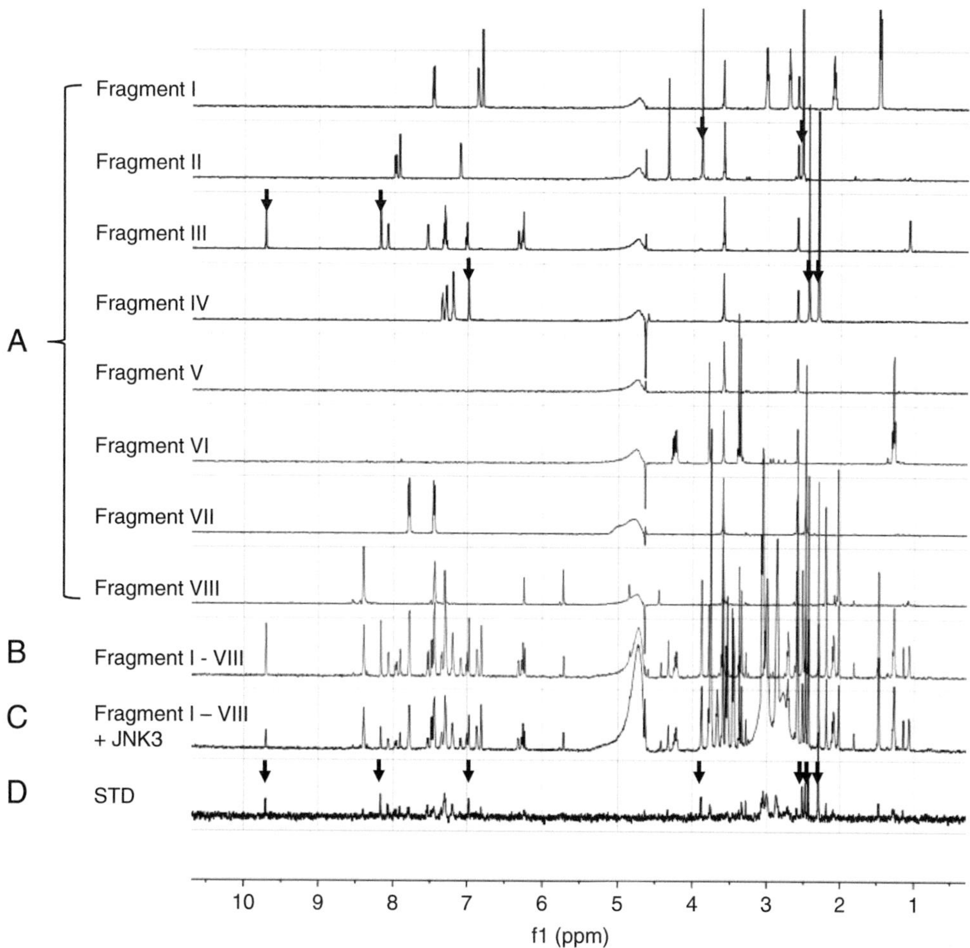

Fig. 2 NMR screening by saturation transfer difference experiments. 1D ^{1}H NMR spectra of fragment I to VIII (A), 8-compound mixture (B), and JNK3 catalytic domain with 8-compound mixture (C). The resonances of JNK3 binders are seen and indicated by *arrow* in STD spectrum (D). The corresponding resonances on individual fragments are also indicated by *arrow*

3.7 Post-screening Steps

Using this protocol, we identified 20 small molecules which are supposed to bind JNK3 catalytic domain. To double-check the binding of those compounds to JNK3, we employed SPR technique. Most of them showed micro- to millimolar affinity to JNK3. We selected 10 compounds for further structure determination by X-ray crystallography and were able to determine two structures of fragment bound JNK3 catalytic domain. Based on the complex structure, the chemical modification of these compounds is undergoing for lead optimization.

4 Notes

1. Highly deuterated DMSO ($_{d6}$-DMSO) is preferable in NMR screening. If the protonated DMSO is used in NMR experiments, there will remain strong ^{1}H resonances from the protonated solvent around 2.5 ppm in NMR spectrum, which hinders the identification of the protein-bound fragment. For the same reason, the use of $_{d11}$-Tris-Cl buffer in 100 % D_2O is necessary.
2. 32 scans gave rise to a high-quality 1D ^{1}H NMR spectrum for 320 μM fragment compound, and the running time was less than 1 min for each sample. If the signal-to-noise ratio is good enough, the concentration of the compound can be reduced. Usually several hundred micro-molar concentration is satisfactory both for the signal-to-noise ratio and total NMR running time.
3. In STD NMR experiments the concentration of the ligand should be kept much higher than that of the receptor. To determine the molar ratio of the ligand to the receptor and to verify the efficiency of STD NMR experiment, it is necessary to run the experiment with a known binder such as JAK3 inhibitor SP600125 which can be used as a positive control molecule. However, since SP600125 has limited water solubility, it is not easy to get a good enough quality NMR spectra at high concentration of the inhibitor. Instead, ATP can be used for this purpose. Keep the concentration of ATP to 160 μM and measure the 1D ^{1}H NMR spectrum. Also run STD NMR measurement with the sampling containing 160 μM ATP and various concentration of JNK3, such as, 5, 10, 20, 50, and 100 μM. We were able to get the STD NMR spectrum with sufficient sensitivity for the sample containing 160 μM ATP and 5 μM JNK3. Discrimination of the receptor binder from non-binder in STD NMR was successful when we added a PCR dNTP mixture into the NMR sample containing 160 μM ATP and 5 μM JNK3. The concentrations of used dNTP mixture were equivalent to that of ATP and only ATP resonances were found in the STD NMR spectrum validating that only ATP binds JAK3 (Fig. 3).
4. During the procedure stated in *see* **Note 3**, the parameters for STD experiment should be optimized. The most important one is the on-resonance saturation frequency of the receptor. Literatures on STD NMR usually recommend -1 or -2 ppm for the saturation frequency; however, saturation on lower magnetic field like 0 or 0.5 ppm sometimes yields more enhanced STD signals. It should be noted that there is no ligand resonances around the saturation frequency. Off-resonance saturation was carried out at 30 ppm where none of the resonances from the receptor or ligand could be found.

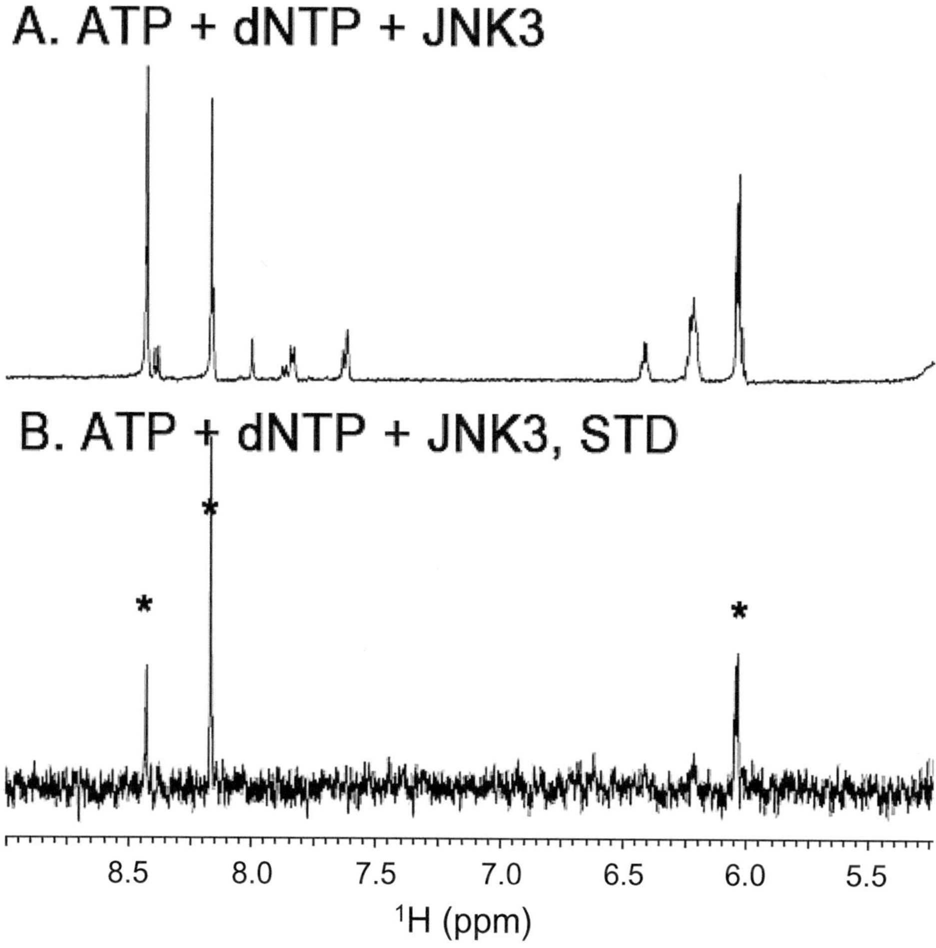

Fig. 3 Optimization of STD NMR experiment. ATP and dNTP mixtures were added to JNK3 catalytic domain and 1D ^{1}H NMR (**a**) and STD NMR (**b**) spectra were recorded. Only resonances of ATP were detected in STD NMR. During these measurement the number of scans, saturation frequency, and the ratio of receptor to compounds should be optimized (*see* Note 4)

Acknowledgement

This work was supported by the GRRC program of Gyeonggi province [(GRRC-DONGGUK2014-B02), diseasesDevelopment of novel recombinant enzymes and diagnostic platform].

References

1. Zhang J, Yang PL, Gray NS (2009) Targeting cancer with small molecule kinase inhibitors. Nat Rev Cancer 9:28–39
2. Cohen P, Alessi DR (2013) Kinase drug discovery—what's next in the field? ACS Chem Biol 8:96–104
3. Shuker SB, Hajduk PJ, Meadows RP, Fesik SW (1996) Discovering high-affinity ligands for proteins: SAR by NMR. Science 274:1531–1534
4. Murray CW, Verdonk ML, Rees DC (2012) Experiences in fragment-based drug discovery. Trends Pharmacol Sci 33:224–232

5. Bollag G, Hirth P, Tsai J et al (2010) Clinical efficacy of a RAF inhibitor needs broad target blockade in BRAF-mutant melanoma. Nature 467:596–599
6. Tsai J, Lee JT, Wang W et al (2008) Discovery of a selective inhibitor of oncogenic B-Raf kinase with potent antimelanoma activity. Proc Natl Acad Sci U S A 105:3041–3046
7. Drysdale MJ, Brough PA (2008) Medicinal chemistry of Hsp90 inhibitors. Curr Top Med Chem 8:859–868
8. Murray CW, Carr MG, Callaghan O et al (2010) Fragment-based drug discovery applied to Hsp90. Discovery of two lead series with high ligand efficiency. J Med Chem 53:5942–5955
9. Woodhead AJ, Angove H, Carr MG et al (2010) Discovery of (2,4-dihydroxy-5-isopropylphenyl)-[5-(4-methylpiperazin-1-ylmethyl)-1,3-dihydroisoindol-2-yl] methanone (AT13387), a novel inhibitor of the molecular chaperone Hsp90 by fragment based drug design. J Med Chem 53:5956–5969
10. Wang YS, Strickland C, Voigt JH et al (2010) Application of fragment-based NMR screening, X-ray crystallography, structure-based design, and focused chemical library design to identify novel microM leads for the development of nM BACE-1 (beta-site APP cleaving enzyme 1) inhibitors. J Med Chem 53:942–950
11. Zhu Z, Sun ZY, Ye Y et al (2010) Discovery of cyclic acylguanidines as highly potent and selective beta-site amyloid cleaving enzyme (BACE) inhibitors: Part I-inhibitor design and validation. J Med Chem 53:951–965
12. Baker M (2013) Fragment-based lead discovery grows up. Nat Rev Drug Discov 12:5–7
13. Campos-Olivas R (2011) NMR screening and hit validation in fragment based drug discovery. Curr Top Med Chem 11:43–67
14. Meyer B, Peters T (2003) NMR spectroscopy techniques for screening and identifying ligand binding to protein receptors. Angew Chem Int Ed Engl 42:864–890
15. Mayer M, Meyer B (2003) Group epitope mapping by saturation transfer difference NMR to identify segments of a ligand in direct contact with a protein receptor. J Am Chem Soc 123:6108–6117
16. Mayer M, Meyer B (1999) Characterization of ligand binding by saturation transfer difference NMR spectroscopy. Angew Chem Int Ed Engl 38:1784–1788
17. Streiff JH, Juranic NO, Macura SI et al (2004) Saturation transfer difference nuclear magnetic resonance spectroscopy as a method for screening proteins for anesthetic binding. Mol Pharmacol 66:929–935
18. Jancarik J, Pufan R, Hong C, Kim SH, Kim R (2004) Optimum solubility (OS) screening: an efficient method to optimize buffer conditions for homogeneity and crystallization of proteins. Acta Crystallogr D Biol Crystallogr 60:1670–1673
19. Lau WF, Withka JM, Hepworth D et al (2011) Design of a multi-purpose fragment screening library using molecular complexity and orthogonal diversity metrics. J Comput Aided Mol Des 25:621–636
20. Probst GD, Bowers S, Sealy JM et al (2011) Highly selective c-Jun N-terminal kinase (JNK) 2 and 3 inhibitors with in vitro CNS-like pharmacokinetic properties prevent neurodegeneration. Bioorg Med Chem Lett 21: 315–319
21. Cee VJ, Cheng AC, Romero K et al (2009) Pyridyl-pyrimidine benzimidazole derivatives as potent, selective, and orally bioavailable inhibitors of Tie-2 kinase. Bioorg Med Chem Lett 19:424–427
22. Bennett BL, Sasaki DT, Murray BW et al (2001) SP600125, an anthrapyrazolone inhibitor of Jun N-terminal kinase. Proc Natl Acad Sci U S A 98:13681–13686

Chapter 4

Bioluminescence Methods for Assaying Kinases in Quantitative High-Throughput Screening (qHTS) Format Applied to Yes1 Tyrosine Kinase, Glucokinase, and PI5P4Kα Lipid Kinase

Mindy I. Davis, Douglas S. Auld, and James Inglese

Abstract

Assays in which the detection of a biological phenomenon is coupled to the production of bioluminescence by luciferase have gained widespread use. As firefly luciferases (FLuc) and kinases share a common substrate (ATP), coupling of a kinase to FLuc allows for the amount of ATP remaining following a kinase reaction to be assessed by quantitating the amount of luminescence produced. Alternatively, the amount of ADP produced by the kinase reaction can be coupled to FLuc through a two-step process. This chapter describes the bioluminescent assays that were developed for three classes of kinases (lipid, protein, and metabolic kinases) and miniaturized to 1536-well format, enabling their use for quantitative high-throughput (qHTS) of small-molecule libraries.

Key words Quantitative high-throughput screening (qHTS), Yes1, Glucokinase, PI5P4Kα, Kinase, Bioluminescence, Luciferase, ADP-Glo

1 Introduction

Kinases are a diverse class of enzymes that phosphorylate a variety of substrates, including proteins, lipids, and metabolites. Dysregulation of phosphorylation leads to diseases such as cancer, inflammation, and diabetes [1]. Indeed, a variety of drugs that target kinases have been FDA approved but as yet there are many kinases that show promise for the development of a targeted therapy but for which there is no FDA-approved drug [1]. Therefore, there is widespread interest in developing inhibitors or activators of kinases as a treatment for disease. Methods to screen kinases in high-throughput formats are important to be able to find leads from chemical libraries and potentially aid the development of kinase-directed therapeutics. For a successful high-throughput screen, both the selected assay methodology and the library design are important.

Hicham Zegzouti and Said A. Goueli (eds.), *Kinase Screening and Profiling: Methods and Protocols*, Methods in Molecular Biology, vol. 1360, DOI 10.1007/978-1-4939-3073-9_4, © Springer Science+Business Media New York 2016

The majority of kinases utilize ATP as the phosphate donor to the substrate and have ADP and the phosphorylated substrate as a product. An important exception is pyruvate kinase which transfers a phosphate from phosphoenolpyruvate (PEP) to ADP yielding ATP as a product [2]. Some of the main ways that compounds are assessed for their impact on kinases include quantitating the phosphorylated product with specific antibodies such as HTRF KinEase (Cisbio) [3], competition binding using technologies such as Kinomescan (DiscoveRx) [4], FP assays such as Transcreener ADP Assay (Bellbrooks) [5], and γ-ATP transfer of radiolabel to product [6]. More recently, kinase assays have been assessed by coupling the kinase reaction to that of FLuc, which utilizes ATP to generate bioluminescence through the production of luciferin [7].

Luciferase-coupled kinase reactions are homogeneous mix-and-read assays that produce a stable luminescent glow with good signal to background, suitable for high-throughput screening (HTS) applications. A direct coupling of an ATP-dependent kinase to FLuc would involve running the kinase reaction to ~50–80 % completion followed by FLuc addition to quantify the remaining substrate ATP in endpoint mode using a reagent such as Kinase-Glo™ (Promega) [8] or EasyLite™ (Perkin Elmer). This high degree of substrate conversion is needed to have sufficient signal to background (S/B) for the development of a robust assay. Ideally, one would like to monitor product formation so that a suitable S/B can be achieved at a much lower level of conversion; generally 5–20 % conversion is sufficient, thereby leading to a more sensitive assay design. However, the theoretical shift in IC_{50} values for an enzyme assay performed at a high percent substrate conversion (e.g., ~80 % conversion) is approximately twofold, which can be acceptable for screening applications [9]. Ideally, initial rate conditions are established in which the enzyme assay has low % conversion to determine accurate enzyme parameters [10]. ADP-Glo™ (Promega) [11, 12] involves a two-step process for ADP quantification, initially depleting the residual substrate ATP through the action of a soluble adenylyl cyclase in a first step, followed by cyclase inhibition and conversion of the ADP from the kinase reaction into ATP by a nucleotide kinase such as pyruvate kinase with the signal generated once again using a "glow-type" of FLuc (Ultra-Glo™ Recombinant Luciferase; *see* 15) to quantitate ATP (*see* Fig. 1) [13]. For a given kinase reaction, the luminescence signal of the Kinase-Glo™-coupled assay increases as the kinase reaction is inhibited, leaving more substrate ATP while for ADP-Glo™-coupled assay the reverse trend occurs (*see* Fig. 2) [7]. As there are compounds that inhibit FLuc [14, 15], it is important to run counterassays to assess detection interference following the identification of hits from a luciferase-coupled primary screen [16]. For example, GSK has released a protein kinase inhibitor set (PKIS)

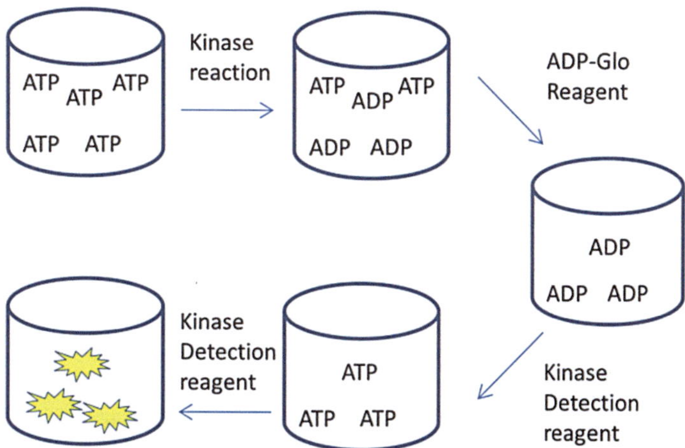

Fig. 1 Schematic of the ADP-Glo™ kit detection process: Following the kinase reaction the remaining ATP is removed by the ADP-Glo™ reagent and then the kinase detection reagent generates ATP from the kinase reaction ADP and uses this ATP to generate luminescence

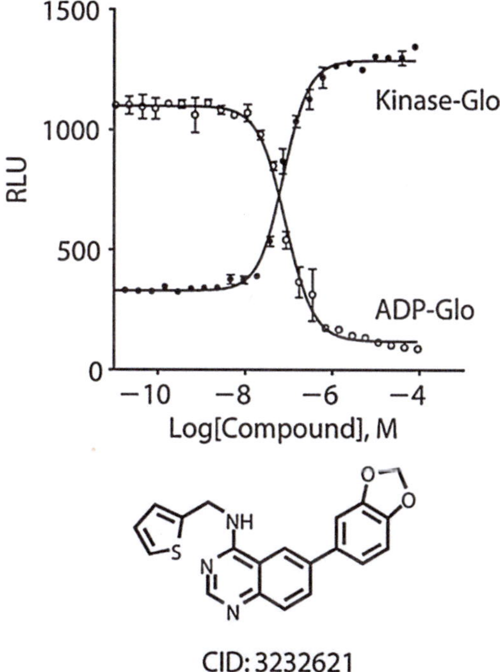

Fig. 2 Measurement of ATP or ADP levels following a kinase reaction: The raw light units (RLUs) are plotted for an inhibitor following a CLK4 assay using the Kinase-Glo™ assay (*solid circles*) and the ADP-Glo™ assay (*open circles*) detection techniques [7]

developed from published kinase inhibitors that has been profiled against firefly and renilla reniformis luciferase to aid in the interpretation of assay results derived from the use of this library [17]. The availability of libraries targeted toward the kinome developed from so-called privileged structures (*see* 7) can greatly facilitate the identification of chemical scaffolds from which lead optimization can progress.

The ADP-Glo™ technology was applied to three classes of kinases here indicating the broad utility of this assay methodology: a lipid kinase PI5P4Kα [18], a tyrosine kinase Yes1 [19], and a metabolic kinase glucokinase (GCK) [20]. The primary goal was to identify inhibitors of PI5P4Kα and Yes1 for the potential use as anticancer agents and activators of GCK for the potential treatment of diabetes. Each of these assays was miniaturized to the 1536-well format and used to screen small-molecule compound libraries in dose response, a method termed quantitative high-throughput screening (qHTS) [21].

2 Materials

Prepare and store solutions at room temperature unless otherwise noted. Prepare all solutions using ultrapure water.

2.1 Yes1 Tyrosine Kinase

1. Yes1 buffer: 6 nM Yes1 (Millipore), 50 mM Tris pH 7.5, 1 mM DTT, 150 mM NaCl, 0.01 % Brij-35, 5 % glycerol. Mix gently and store at room temperature. For continuous dispensing of multiple plates store Yes1 buffer at 4 °C (*see* **Note 1**).
2. Assay buffer: 50 mM Tris pH 7.5, 1 mM DTT, 150 mM NaCl, 0.01 % Brij-35, 5 % glycerol. Mix gently and store at room temperature.
3. Substrate Poly(E_4Y) (Sigma) was dissolved in 50 mM Tris pH 7.5, 150 mM NaCl, and 5 % glycerol at 10 mg/mL, aliquoted, and stored at −80 °C.
4. Yes1 substrate buffer: 0.3 mM ATP, 0.3 mM EGTA, 30 mM $MgCl_2$, 0.9 mg/mL poly(E_4Y), 50 mM Tris pH 7.5, 1 mM DTT, 150 mM NaCl, 0.01 % Brij-35, 5 % glycerol. Mix gently and store at room temperature. The Ultrapure ATP from the ADP-Glo™ kit was used (*see* **Notes 2** and **3**).
5. ADP-Glo™ Reagent: Thaw at room temperature per the manufacturer's protocol.
6. Kinase detection reagent: Thaw at room temperature per the manufacturer's protocol. Add kinase detection buffer to powdered kinase detection substrate and swirl gently until fully dissolved.

2.2 Glucokinase

1. GCK buffer: 22.5 nM GCK, 22.5 nM GCK regulatory protein, 3 mM $MgCl_2$, 37.5 mM KCl, 37.5 mM Hepes, 1.5 mM DTT, 0.0375 % BSA, 0.015 % Tween-20, pH adjusted to 7.1. Proteins were expressed and purified as described in [20] (*see* **Note 1**).
2. No GCK buffer: 22.5 nM GCK regulatory protein, 3 mM $MgCl_2$, 37.5 mM KCl, 37.5 mM Hepes, 1.5 mM DTT, 0.0375 % BSA, 0.015 % Tween-20, pH adjusted to 7.1.
3. GCK substrate buffer: 1.2 mM ATP and 15 mM glucose in water. The Ultrapure ATP from the ADP-Glo™ kit was used (*see* **Notes 2** and **3**).
4. ADP-Glo™ Reagent: Thaw at room temperature per the manufacturer's protocol.
5. Kinase detection reagent: Thaw at room temperature per the manufacturer's protocol. Add kinase detection buffer to powdered kinase detection substrate and swirl gently until fully dissolved.

2.3 Lipid Preparation

1. DPPS (1,2-dipalmitoyl-sn-glycero-3-phosphoserine) (Echelon Biosciences) was suspended in DMSO (333 μL DMSO per 1 mg DPPS), sonicated for 1 min, and mixed by vortexing for 30 s, forming a solution (*see* **Note 4**).
2. PI5P (D-myo-phosphatidylinositol 5 phosphate diC16) from Echelon Biosciences was suspended in DMSO and mixed by vortexing for several minutes (333 μL DMSO per 1 mg PI5P).
3. Lipid mix: DPPS:PI5P (2:1 ratio), made by mixing 1000 μL of DPPS to 500 μL of PI5P.
4. 1500 μL of DMSO was added and the resulting lipid mixture was alternately sonicated and vortexed for several minutes. The result is a suspension with no visible particulate matter (*see* **Notes 5–8**).

2.4 PI5P4Kα Lipid Kinase

1. PI5P4Kα/PI5P buffer: 15 nM PI5P4Kα, 112.5 μM PI5P (Echelon Biosciences), 225 μM DPPS (Echelon Biosciences), 40 mM Hepes pH 7.4, 0.25 mM EGTA, 0.1 % CHAPS. Protein was expressed and purified as described in [18] (*see* **Note 1**). To make this reagent, 1259 μL of lipid mix described in Subheading 2.3 was added to 315 μL of DMSO. Then 5287 μL of buffer 1 (30 mM Hepes pH 7.4, 1 mM EGTA, 0.1 % CHAPS) was added and the mixture was sonicated. Then, 13525 μL of buffer 2 (50 mM Hepes pH 7.4, 0.1 % Chaps) was added and the mixture was sonicated. Lastly, enzyme was added and the solution was gently mixed by pipetting. This reagent was stored on wet ice.
2. No PI5P4Kα buffer: 112.5 μM PI5P (Echelon Biosciences), 225 μM DPPS (Echelon Biosciences), Hepes pH 7.4, EGTA, 0.1 % CHAPS.

3. No PI5P buffer: 15 nM PI5P4Kα, Hepes pH 7.4, EGTA, 0.1 % CHAPS.
4. ATP buffer: 15 μM ATP, 20 mM Hepes 7.4, 60 mM $MgCl_2$, 0.1 % CHAPS (*see* **Notes 2** and **3**).
5. ADP-Glo™ Reagent: Thaw at room temperature per the manufacturer's protocol (*see* **Note 9**).
6. Kinase detection reagent: Thaw at room temperature per the manufacturer's protocol. Add kinase detection buffer to powdered kinase detection reagent and swirl gently until fully dissolved (*see* **Notes 10–12**).

3 Methods

3.1 Yes1 Tyrosine Kinase

1. Dispense 2 μL of Yes1 enzyme buffer into all but one column of a white 1536-well solid bottom plate (Greiner) with a Flying Reagent Dispenser (Beckman Coulter) leaving a control column that will contain just the buffer components but no Yes1 (*see* **Note 13**).
2. Dispense 2 μL of the assay buffer control in the remaining column as a no-enzyme control.
3. Transfer 23 nL of library compounds dissolved in DMSO from a 1536-well compound plate arrayed in columns 5–48 into the assay plate using a pintool (Kalypsys Systems) [22] (*see* **Note 4**).
4. Transfer 23 nL of control compounds (DMSO used here in column 1, 3–4 and known Yes1 inhibitor dasatinib [23] in 16-pt dose–response in column 2) from a 1536-well compound plate arrayed in columns 1–4 into the assay plate using the pintool.
5. Incubate the enzyme with compound for 15 min.
6. Initiate the enzyme reaction by dispensing 1 μL of Yes1 substrate buffer to all wells of the assay plate.
7. Cover the plate with a solid lid.
8. Incubate at room temperature for 1 h.
9. Add 2 μL of ADP-Glo™ reagent (*see* **Note 14**).
10. Incubate the lidded plate for 40 min.
11. Add 4 μL of kinase detection reagent (*see* **Note 15**).
12. Incubate the lidded plate for 30 min (*see* **Note 16**).
13. Read the luminescence signal with a ViewLux (Perkin-Elmer) with a 1-s exposure.
14. Normalize the data to DMSO-treated control columns with (maximum signal) and without (minimum signal) enzyme (*see* **Note 17**).

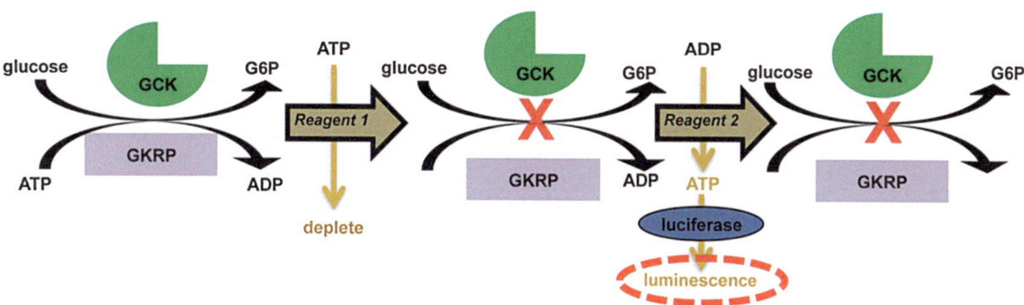

Fig. 3 Schematic of the ADP-Glo™ assay for glucokinase: This assay monitors glucokinase activity directly but also indirectly monitors the protein–protein interaction of glucokinase with glucokinase regulatory protein. A small molecule that disrupts the interaction of glucokinase with glucokinase regulatory protein would lead to an increase in observed luminescence [20]

3.2 Glucokinase

The GCK assay was designed to look for inhibitors of the GCK/GCK regulatory protein interaction which would lead to activation of GCK (*see* Fig. 3). This assay could be run with just GCK to look for activators or inhibitors. Additional complementary assay formats for interrogating the interaction of GCK with GCK regulatory protein can be found in [20, 27].

1. Dispense 2 μL of GCK/GCK regulatory enzyme buffer into all but one column of a white 1536-well solid bottom plate (Greiner) with a Flying Reagent Dispenser (Beckman Coulter) leaving a control column that contains just the buffer components but no GCK (*see* **Note 13**).
2. Dispense 2 μL of no GCK buffer control in the remaining column.
3. Transfer 23 nL of library compounds dissolved in DMSO from a 1536-well compound plate arrayed in columns 5–48 into the assay plate using a pintool (Kalypsys Systems).
4. Transfer 23 nL of control compounds (DMSO used here in columns 1, 3–4 and known GCK activator cmpd A from EMD Millipore (GKA-EMD) in 16-pt dose–response) from a 1536-well compound plate arrayed in columns 1–4 into the assay plate using the pintool.
5. Incubate the enzyme with compound for 15 min.
6. Initiate the enzyme reaction by dispensing 1 μL of GCK substrate buffer to all wells of the assay plate.
7. Cover the plate with a solid lid.
8. Incubate at room temperature for 1 h.
9. Add 2.5 μL of ADP-Glo™ reagent.
10. Incubate the lidded plate for 40 min.
11. Add 5 μL of kinase detection reagent.
12. Incubate the lidded plate for 30 min.

13. Read the luminescence signal with a ViewLux with a 1-s exposure.
14. Normalize the data to DMSO-treated control columns with (maximum signal) and without (minimum signal) enzyme.

3.3 PI5P4Kα Lipid Kinase

1. Dispense 2 μL of PI5P4Kα/PI5P enzyme buffer into all but two columns of a white 1536-well solid bottom plate (Greiner) with a Flying Reagent Dispenser (Beckman Coulter) leaving a control column that contains just the buffer components but no PI5P4Kα and a control column that contains just the buffer components but no PI5P (*see* **Notes 13** and **18**).
2. Dispense 2 μL of no PI5P4Kα buffer control and no PI5P buffer control in the remaining columns.
3. Centrifuge for 10 s at 300 × *g*.
4. Transfer 23 nL of library compounds dissolved in DMSO from a 1536-well compound plate arrayed in columns 5–48 into the assay plate using a pintool (Kalypsys Systems).
5. Transfer 23 nL of control compounds (DMSO used here in columns 1, 3–4 and known PI5P4Kα inhibitor Tyrphostin AG82 (Cayman Chemical Company) [18] in 16-pt dose–response) from a 1536-well compound plate arrayed in columns 1–4 into the assay plate using the pintool.
6. Incubate the enzyme with compound for 15 min.
7. Initiate the enzyme reaction by dispensing 1 μL of ATP buffer to all wells of the assay plate.
8. Cover the plate with a solid lid.
9. Incubate at room temperature for 1 h.
10. Add 2 μL of ADP-Glo™ reagent.
11. Incubate the lidded plate for 40 min.
12. Add 4 μL of kinase detection reagent.
13. Incubate the lidded plate for 30 min.
14. Read the luminescence signal with a ViewLux (Perkin-Elmer) with a 20-s exposure (*see* **Note 19**).
15. Normalize the data to DMSO-treated control columns with (maximum signal) and without (minimum signal) enzyme.

3.4 Counterscreen, Mechanism of Action, and Assay Statistics

FLuc (Sigma cat # L9506) can be used as a counterscreen [14, 16] but for ADP-Glo™ assays it is preferable to use the kit itself and not just FLuc enzyme as the ADP-Glo™ system contains additional proprietary enzyme components. By testing the compound in the presence of the ATP/ADP mix representing the % conversion attained in the assay, it can be determined whether the compound interferes with any of the components of the ADP-Glo™ detection system, including FLuc.

1. For the Yes1 assay, the average $Z' = 0.76$, CV = 6.9 %, and S/B = 23.7 [19]. IC_{50} for control dasatinib = 0.5 nM. Data were deposited in PubChem AID 686947 (primary assay) and 686950 (detection counterassay).
2. For the GCK assay, the average $Z' = 0.7$, CV = 4.2 %, and S/B = 4.2 [20]. Data were deposited in PubChem AID 743206.
3. For the PI5P4Kα assay, the $Z' = 0.77$, CV = 9.3 %, and S/B = 12.6 [18]. Data were deposited in PubChem AID 652105, 652103, 743286, and 743285 (detection counterassay).
4. These assay formats can be used to assess mechanism of inhibition by varying the [ATP] relative to the K_m as long as the amount of ATP used at each point lies within the linear range of the detection reagent ~20 nM to 1 mM as was done previously for PI5P4Kα [18]. Competitive inhibitors will show increased IC_{50} values with increasing concentrations of ATP, noncompetitive inhibitors will show no change, and uncompetitive inhibitors will have decreased IC_{50} values with increasing concentration of ATP.

4 Notes

1. The kinase enzymes and protein substrates are stored at −80 °C and then stored on wet ice until ready to use. The substrates are stored at −20 °C and stored at room temperature until ready to use.
2. The reactions described herein were run with ATP near the K_m. This allows for maximum sensitivity at identifying inhibitors that are competitive, uncompetitive, or noncompetitive with respect to ATP [24]. If the K_m is unknown for your kinase, the coupled pyruvate kinase/lactate dehydrogenase method can be used to determine it [25].
3. The presence of contaminating ADP in the ATP used for the kinase reaction will limit the attainable signal to background, so the use of Ultrapure ATP, such as the ATP present in the ADP-Glo™ kit, is recommended.
4. Test compounds are dissolved in DMSO. The tolerability of the kinase to DMSO should be tested. PI5P4Kα was sensitive to DMSO levels higher than 7.25 % [18]. The assays described herein use 5 % DMSO, which is needed to solubilize the PI5P lipid. The remaining assays, GCK and Yes1 kinase, had <1 % DMSO as DMSO was needed only to solubilize the compounds and not the substrates in those assays.
5. Lipid solutions should be prepared and stored in glass containers to minimize the lipid sticking to plastic.

6. Lipid prep of PI5P and DPPS in DMSO can be stored at −20 °C and thawed and sonicated prior to use. The PI5P/DPPS mixture is stable to at least six freeze/thaw cycles.
7. PI5P/DPPS is a suspension and is slightly opaque. The PI5P4Kα/PI5P reagent was stable for at least 16 h making it amenable to HTS and is a solution with a CV and *Z*′ across each 1536-well plate indicating that the delivery of the lipid mixture is quite uniform. PI5P alone is minimally soluble in DMSO, so the DPPS DMSO solution can be added to the PI5P followed by sonication to assist the formation of the uniform suspension. An alternate assay method for PI5P4Kα using liposomes is described in [26].
8. While there are shorter chain PI5P lipids available that are more soluble in DMSO, these are not substrates for PI5P4Kα.
9. ADP-Glo™ reagents can be refrozen and stored at −20 °C. Upon thawing mix gently and check for precipitation. Solution can be warmed at 37 °C for 15 min to dissolve precipitate or supernatant can be removed and used.
10. ADP-Glo™ kit reagents and kinase reaction mixture should be at room temperature prior to the detection steps.
11. The ADP-Glo™ kit can be used for any enzyme that generates ADP, including ATPases and kinases.
12. The ADP-Glo™ kit can detect up to 1 mM of ATP/ADP and can also detect as little as 20 nM ADP.
13. It is important to initially have a control that is all components but no enzyme and a control that is all components but no substrate to assess the level of substrate-independent ATPase activity as well as ensure that the substrate does not itself contain a contaminating ATPase.
14. The ADP-Glo™ kit requires the presence of at least 0.5 mM $MgCl_2$, so if the kinase reaction does not have $MgCl_2$ present it should be added with the ADP-Glo™ reagent. Termination of the kinase reaction is not required with the kit and the use of EDTA to terminate the kinase reaction should be avoided as the Mg^{2+} is needed for the ADP-Glo™ kit.
15. The manufacturer's recommendations are to add a 1:1 ratio of ADP-Glo™ reagent to your kinase reaction and then a 1:1 ratio of kinase detection reagent to that. The minimum volume that is preferable in a 1536-well plate is 3 μL but the maximum volume the plate can hold is 12 μL with a preference for staying at or below 10.5 μL, so the ADP-Glo™ kit reagents were used at levels below the manufacturer's recommendation. For the assays used herein, this did not adversely affect the assay.
16. The ADP-Glo™ signal/background is stable for at least 3 h per the manufacturer's specifications making it amenable to HTS.

17. A standard curve can be made using the starting amount of ATP in the kinase reaction and creating admixtures of ATP/ADP representing 0–100 % conversion.
18. The bottles used to dispense reagents on the Flying Reagent Dispenser are made of plastic. For the PI5P4Kα/PI5P reagent a glass test tube was placed inside the plastic bottle to minimize sticking of the lipid to plastic.
19. The exposure time in the ViewLux is significantly longer for the PI5P4Kα assay than the GCK and Yes1 assays.

Acknowledgement

This work was supported by the Molecular Libraries Initiative of the NIH Roadmap for Medical Research. The content of this publication does not necessarily reflect the views of policies of the Department of Health and Human Services, nor does mention of trade names, commercial products, or organizations imply endorsement by the US Government.

References

1. Cohen P, Alessi DR (2012) Kinase drug discovery—what's next in the field? ACS Chem Biol 8:96–104
2. Walsh MJ, Brimacombe KR, Veith H et al (2011)2-Oxo-N-aryl-1,2,3,4-tetrahydroquinoline-6-sulfonamides as activators of the tumor cell specific M2 isoform of pyruvate kinase. Bioorg Med Chem Lett 21:6322–6327
3. Harbert C, Marshall J, Soh S, Steger K (2008) Development of a HTRF kinase assay for determination of Syk activity. Curr Chem Genomics 1:20–26
4. Fabian MA, Biggs WH 3rd, Treiber DK et al (2005) A small molecule-kinase interaction map for clinical kinase inhibitors. Nat Biotechnol 23:329–336
5. Kleman-Leyer KM, Klink TA, Kopp AL et al (2009) Characterization and optimization of a red-shifted fluorescence polarization ADP detection assay. Assay Drug Dev Technol 7:56–67
6. Hastie CJ, McLauchlan HJ, Cohen P (2006) Assay of protein kinases using radiolabeled ATP: a protocol. Nat Protoc 1:968–971
7. Tanega C, Shen M, Mott BT, Thomas CJ, MacArthur R, Inglese J, Auld DS (2009) Comparison of bioluminescent kinase assays using substrate depletion and product formation. Assay Drug Dev Technol 7:606–614
8. Somberg R, Goueli SA (2004) Method for detecting transferase enzymatic activity. In *US 2004/0101922 A1* (USPTO ed., Promega Corporation, USA
9. Wu G, Yuan Y, Hodge CN (2003) Determining appropriate substrate conversion for enzymatic assays in high-throughput screening. J Biomol Screen 8:694–700
10. Segel IH (1993) Enzyme kinetics: behavior and analysis of rapid equilibrium and steady-state enzyme systems. John Wiley and Sons, New York
11. Li H, Totoritis RD, Lor LA, Schwartz B, Caprioli P, Jurewicz AJ, Zhang G (2009) Evaluation of an antibody-free ADP detection assay: ADP-Glo. Assay Drug Dev Technol 7: 598–605
12. Sanghera J, Li R, Yan J (2009) Comparison of the luminescent ADP-Glo assay to a standard radiometric assay for measurement of protein kinase activity. Assay Drug Dev Technol 7: 615–622
13. Zegzouti H, Goueli SA (2010) ADP Detection based luminescent phosphotransferase or ATP

hydrolase assay. In *US 2010/0075350 A1* (USPTO ed., Promega Corporation, USA

14. Auld DS, Southall NT, Jadhav A, Johnson RL, Diller DJ, Simeonov A, Austin CP, Inglese J (2008) Characterization of chemical libraries for luciferase inhibitory activity. J Med Chem 51:2372–2386
15. Auld DS, Zhang YQ, Southall NT, Rai G, Landsman M, MacLure J, Langevin D, Thomas CJ, Austin CP, Inglese J (2009) A basis for reduced chemical library inhibition of firefly luciferase obtained from directed evolution. J Med Chem 52:1450–1458
16. Thorne N, Shen M, Lea WA, Simeonov A, Lovell S, Auld DS, Inglese J (2012) Firefly luciferase in chemical biology: a compendium of inhibitors, mechanistic evaluation of chemotypes, and suggested use as a reporter. Chem Biol 19:1060–1072
17. Dranchak P, MacArthur R, Guha R, Zuercher WJ, Drewry DH, Auld DS, Inglese J (2013) Profile of the GSK published protein kinase inhibitor set across ATP-dependent and-independent luciferases: implications for reporter-gene assays. PLoS One 8, e57888
18. Davis MI, Sasaki AT, Shen M, Emerling BM et al (2013) A homogeneous, high-throughput assay for phosphatidylinositol 5-phosphate 4-kinase with a novel, rapid substrate preparation. PLoS One 8, e54127
19. Patel PR, Sun H, Li SQ, Shen M, Khan J, Thomas CJ, Davis MI (2013) Identification of potent Yes1 kinase inhibitors using a library screening approach. Bioorg Med Chem Lett 23:4398–4403
20. Rees MG, Davis MI, Shen M, Titus S, Raimondo A, Barrett A, Gloyn AL, Collins FS, Simeonov A (2014) A panel of diverse assays to interrogate the interaction between glucokinase and glucokinase regulatory protein, two vital proteins in human disease. PLoS One 9, e89335
21. Inglese J, Auld DS, Jadhav A, Johnson RL, Simeonov A, Yasgar A, Zheng W, Austin CP (2006) Quantitative high-throughput screening: a titration-based approach that efficiently identifies biological activities in large chemical libraries. Proc Natl Acad Sci U S A 103: 11473–11478
22. Yasgar A, Shinn P, Jadhav A, Auld D, Michael S, Zheng W, Austin CP, Inglese J, Simeonov A (2008) Compound management for quantitative high-throughput screening. JALA Charlottesv Va 13:79–89
23. Rix U, Hantschel O, Durnberger G, Remsing Rix LL et al (2007) Chemical proteomic profiles of the BCR-ABL inhibitors imatinib, nilotinib, and dasatinib reveal novel kinase and nonkinase targets. Blood 110:4055–4063
24. Copeland RA (2005) Evaluation of enzyme inhibitors in drug discovery. A guide for medicinal chemists and pharmacologists. Methods Biochem Anal 46:1–265
25. Jenkins WT (1991) The pyruvate kinase-coupled assay for ATPases: a critical analysis. Anal Biochem 194:136–139
26. Demian DJ, Clugston SL, Foster MM, Rameh L, Sarkes D, Townson SA, Yang L, Zhang M, Charlton ME (2009) High-throughput, cell-free, liposome-based approach for assessing in vitro activity of lipid kinases. J Biomol Screen 14:838–844
27. Lloyd DJ, St Jean DJ Jr, Kurzeja RJ, Wahl RC et al (2013) Antidiabetic effects of glucokinase regulatory protein small-molecule disruptors. Nature 504:437–440

Chapter 5

Using Bioluminescent Kinase Profiling Strips to Identify Kinase Inhibitor Selectivity and Promiscuity

Hicham Zegzouti, Jacquelyn Hennek, and Said A. Goueli

Abstract

The advancement of a kinase inhibitor throughout drug discovery and development is predicated upon its selectivity towards the target of interest. Thus, profiling the compound against a broad panel of kinases is important for providing a better understanding of its activity and for obviating any off-target activities that can result in undesirable consequences. To assess the selectivity and potency of an inhibitor against multiple kinases, it is desirable to use a universal assay that can monitor the activity of all classes of kinases regardless of the nature of their substrates. The luminescent ADP-Glo kinase assay is a universal platform that measures kinase activity by quantifying the amount of the common kinase reaction product ADP. Here we present a method using standardized kinase profiling systems for inhibitor profiling studies based on ADP detection by luminescence. The kinase profiling systems are sets of kinases organized by family, presented in multi-tube strips containing eight enzymes, each with corresponding substrate strips, and standardized for optimal kinase activity. We show that using the kinase profiling strips we could quickly and easily generate multiple selectivity profiles using small or large kinase panels, and identify compound promiscuity within the kinome.

Key words Kinase profiling, Bioluminescence, ADP detection, Selectivity profiles, Kinase assay

1 Introduction

Kinases are phosphotransferases that account for one of the largest enzyme families in the cell. By phosphorylating a multitude of substrates, kinases transduce cellular signals altering diverse biological networks [1, 2]. Because of their importance, alterations in normal kinase activity (e.g., overexpression, hyper-activation, or inhibition) can disrupt several of these networks causing diseases such as cancer, inflammation, and diabetes. Thus, kinases became one of the largest drug target enzyme groups in the drug discovery research [3].

Many drug discovery programs are devoted to the identification of kinase inhibitors. Typically, these programs consist of screening small molecule libraries of different sizes and chemical

Hicham Zegzouti and Said A. Goueli (eds.), *Kinase Screening and Profiling: Methods and Protocols*, Methods in Molecular Biology, vol. 1360, DOI 10.1007/978-1-4939-3073-9_5,

structures, to identify compounds that inhibit the target kinase. After screening and hit identification, selected compounds are moved forward for lead optimization. During this phase, compounds are optimized for potency and selectivity to improve efficacy and prevent off-target effects once the compounds become a drug. Drug safety is of paramount importance, indicating that minimal side effects are a major requirement in drug development. Because of the large number of kinases with sequence similarity in their catalytic domain, dialing in selectivity is challenging, and off-target kinase inhibition can be a significant source of side effects including undesirable toxicities [4, 5]. Therefore, to find a balance between potency and selectivity, lead compounds need to be profiled against various liability panels, which include a protein kinase panel. Profiling a compound against a broad panel of kinases reveals its activity against off-target kinases, and this may provide insights about potential toxicities as well as mode of action [6]. Moreover, defining specificity of inhibitors towards a specific kinase or group of kinases provides useful tools to the research community in general and to the cell-signaling community in particular [7].

Several detection technologies were developed for assessing kinase activity [8–10]. These can generate kinome selectivity data and in a profiling mode they have mostly been deployed in a fee-for-service model by service providers [11–14]. This is an attractive model given the cost and complexity of setting up profiling panels, which involves rigorous optimization of each enzyme across the panel [15–17]. On the other hand, a profiling system that can be simply and rapidly implemented in-house would obviate logistical inconveniences, delays, and confidentiality concerns associated with outsourcing. To this end, a pre-configured kinase profiling system was developed based on the luminescent ADP-Glo kinase assay platform [9]. This universal kinase platform has been validated with hundreds of kinases and it has been shown to be robust and suitable for kinase drug discovery [18–20]. The kinase profiling systems are a set of kinases presented in multi-tube strips containing eight enzymes per strip with their corresponding substrate strips, and standardized for optimal kinase activity (Fig. 1). The strip system provides flexible kinase inhibitor profiling, as each strip can be used to profile compounds at a single dose or create a dose–response against eight kinases at once. We showed that using this system we could generate selectivity profiles using small or large kinase panels, and identify compound promiscuity towards members of a single kinase subfamily or different subfamilies of the kinome [21]. As an example here, we used a strip containing eight different recombinant receptor tyrosine kinases to profile eight known kinase inhibitors. We generated single-dose and dose–response inhibition data, and confirmed that the inhibition potencies observed with this approach are consistent with published values produced by other technologies [12].

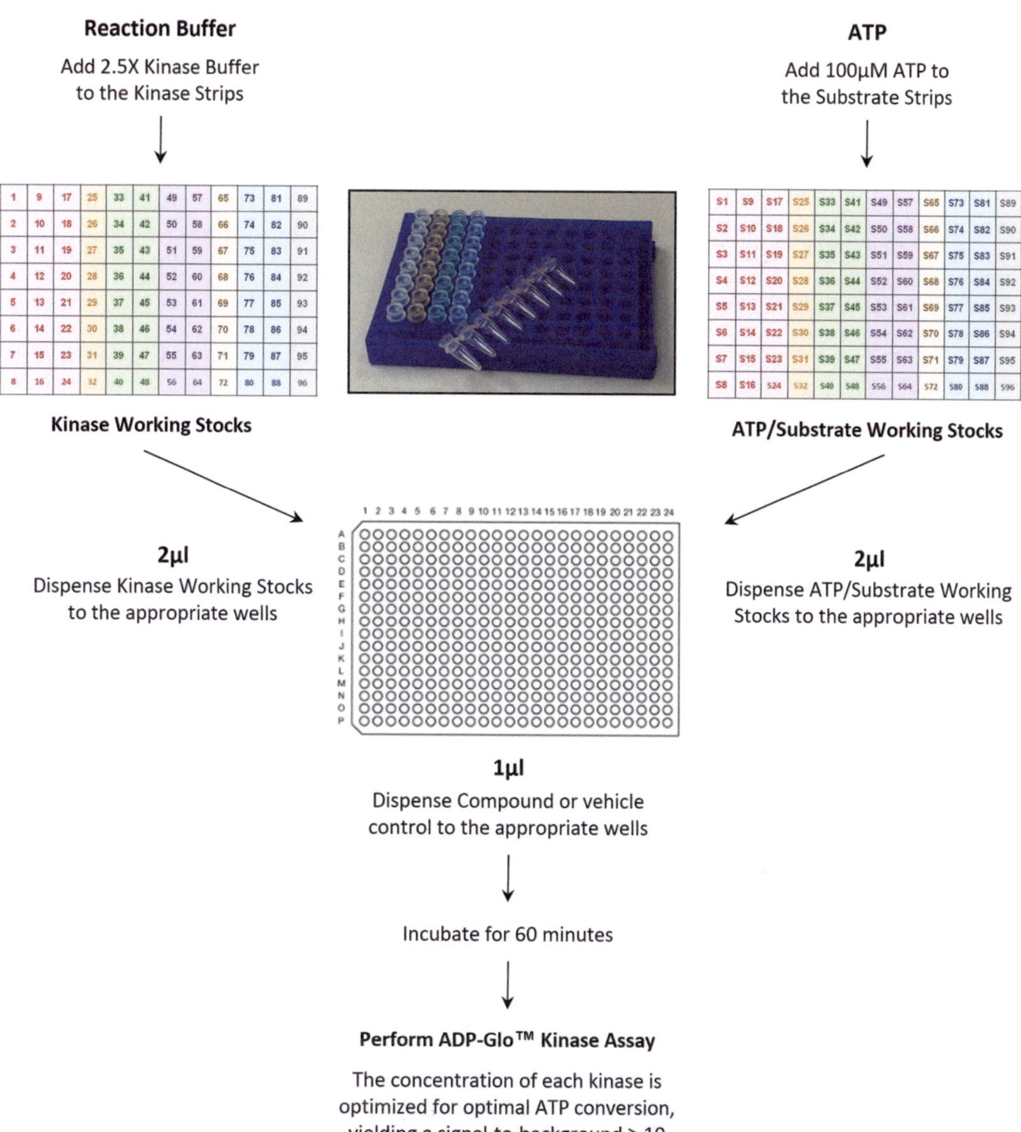

Fig. 1 Kinase selectivity profiling system overview: Two dilutions are used to create kinase working stocks and ATP/substrate working stocks. 2 μL of these stocks are added to a 384-well plate containing 1 μL of test compounds/vehicle to start kinase profiling experiments. Kinase activity in each well is quantified using ADP-Glo™ Kinase Assay. The luminescent signal generated by the ADP-Glo™ Kinase Assay is proportional to ADP concentration and is correlated with kinase activity

2 Materials

Prepare all solutions using Nanopure water, or equivalent water free of ATP. Follow all temperature-controlled steps as noted.

2.1 Single-Dose and Dose–Response Profiling

For one pair of kinase/substrate strips in the kit, up to ten compounds can be tested in a single-dose format or one compound in a dose–response format. This includes duplicate measurements per compound, no-compound vehicle controls, and no-kinase controls for each kinase in the strip. We tested one strip of eight kinases with eight compounds at 1 μM final concentration. Additionally, we used two strips to profile two compounds in dose–response mode.

1. Two kinase selectivity profiling systems (KSPS): TK-1 kits (Promega). One Kit includes:
   (a) 5× Reaction buffer A: 200 mM Tris–HCl, pH 7.5; 100 mM $MgCl_2$; 0.5 mg/mL BSA.
   (b) 0.1 M dithiothreitol (DTT) in water.
   (c) Two 8-tube strips containing stock substrate/cofactor solutions.
   (d) Two 8-tube strips containing stock kinase solutions.

See Table 1 for KSPS: TK-1 kinase and substrate components.

2. Dimethyl sulfoxide (DMSO) (Sigma).
3. Reaction buffer (2.5×) for *single-dose*profiling: 1.4 mL of 2.5× reaction buffer made with 125 μM DTT just prior to use. Mix 700 μL 5× reaction buffer A, 1.75 μL 0.1 M DTT, and 698.25 μL Nanopure water.
4. Reaction buffer (1×) with 5 % DMSOfor *single-dose* profiling: 100 μL was made by mixing 40 μL reaction buffer (2.5×), 5 μL DMSO, and 55 μL Nanopure water (*see* **Note 1**).

Table 1
Kinase and substrate strip compositions of the kinase selectivity profiling system TK-1

TK-1		
Tyrosine kinase family		
	Kinase strip	**Substrate strip**
A	EGFR	Poly (Glu_4, Tyr_1) + $MnCl_2$
B	HER2	Poly (Glu_4, Tyr_1)
C	HER4	Poly (Glu_4, Tyr_1) + $MnCl_2$
D	IGF1R	IGF1Rtide + $MnCl_2$
E	InsR	Axltide + $MnCl_2$
F	KDR	Poly (Glu_4, Tyr_1)
G	PDGFRα	Poly (Glu_4, Tyr_1)
H	PDGFRβ	Poly (Glu_4, Tyr_1)

5. Reaction buffer (2.5×) for *dose–response* profiling: 3 mL of 2.5× reaction buffer made with 125 μM DTT just prior to use. Mix 1.5 mL 5× reaction buffer A, 3.75 μL 0.1 M DTT, and 1496.25 μL Nanopure water.
6. Reaction buffer (1×) with 5 % DMSO for *dose–response* profiling: 2 mL was made by mixing 800 μL reaction buffer (2.5×), 100 μL DMSO, and 1.1 mL Nanopure water (*see* **Note 1**).
7. SB 203580 compound: 100 μM SB 203580 stock (Promega) in DMSO.
8. Enzastaurin compound: 100 μM Enzastaurin stock (LC Laboratories) in DMSO.
9. PF-477736 compound: 1 mM and 100 μM PF-477736 stocks (Sigma) in DMSO.
10. Gefitinib compound: 1 mM and 100 μM Gefitinib stock (LC Laboratories) in DMSO.
11. Roscovitine compound: 100 μM Roscovitine stock (LC Laboratories) in DMSO.
12. Dasatinib compound: 100 μM Dasatinib stock (LC Laboratories) in DMSO.
13. Tofacitinib compound: 100 μM Tofacitinib stock (LC Laboratories) in DMSO.
14. Tozasertib compound: 100 μM Tozasertib stock (LC Laboratories) in DMSO.
15. ADP-Glo™ Kinase Assay (Promega). Kit includes:
    (a) ADP-Glo™ Reagent.
    (b) Kinase detection reagent.
    (c) 10 mM UltraPure ATP (*see* **Note 2**).

2.2 Equipment

1. Low-volume 384-well white round-bottom untreated polystyrene microplate (Corning).
2. Half area 96-well white untreated microplate (Corning).
3. Multichannel pipette capable of dispensing 1 and 2 μL volumes.
4. Luminometer capable of reading multi-well plates (GloMax Discover, Promega).
5. Plate shaker.
6. Centrifuge compatible with a 384-well plate.
7. Incubator or plate warmer at 22–25 °C.

2.3 Data Processing

Data processed using Kinase Selectivity Profiling System Data Analysis Worksheets on Promega Tools webpage (http://www.promega.com/resources/tools/kinase-selectivity-profiling-systems-data-analysis-worksheets/).

3 Methods

3.1 Kinase Reaction Setups

3.1.1 Single-Dose Profiling

The dispensing steps described in this section are shown in Fig. 2.

1. Prepare compound solutions for single-dose profiling using 100 μM compound stocks in DMSO for the following compounds: gefitinib, dasatinib, SB 203580, tozasertib, tofacitinib, roscovitine, PF-477736, and enzastaurin. Mix 40 μL reaction buffer (2.5×), 5 μL 100 μM compound stock in DMSO, and 55 μL Nanopure water (*see* **Note 3**).
2. Dilute 5 μL 10 mM UltraPure ATP with 495 μL of Nanopure water to make 100 μM ATP solution.
3. Thaw one substrate/cofactor strip (TK-S1) from the kit on ice.
4. Dilute the TK-S1 strip to make 2.5× ATP/substrate working stocks. Add 15 μL 100 μM ATP solution to each of the tubes in the substrate/cofactor strip. Mix by pipetting up and down several times. Keep the strip on ice until use.
5. Dispense 1 μL reaction buffer (1×) with 5 % DMSO into columns 1 and 2 of the 384-well assay plate (*see* **Note 4**).
6. Add test compounds to 384-well assay plate. Dispense 1 μL of gefitinib solution into each well of column 3. Dispense 1 μL of dasatinib solution into each well of column 4. Dispense 1 μL of SB 203580 solution into each well of column 5. Dispense 1 μL of tozasertib solution into each well of column 6. Dispense 1 μL of tofacitinib solution into each well of column 7. Dispense 1 μL of roscovitine solution into each well of column 8. Dispense 1 μL of PF-477736 solution into each well of column 9. Dispense 1 μL of enzastaurin solution into each well of column 10.
7. Thaw the eight-tube strip containing kinase stocks (TK-1) on ice (*see* **Note 5**).
8. Make 2.5× kinase working stocks by diluting the TK-1 strip. Add 95 μL 2.5× reaction buffer to each tube of the strip and mix by gently pipetting up and down several times. Perform this step immediately prior to use.
9. Add 2 μL 2.5× reaction buffer into column 1 of the 384-well assay plate for no-kinase controls.
10. Align the strip with kinase working stocks, so the blue dye is at the top relative to the 384-well plate (*see* **Note 6**).
11. Using a multichannel pipette (eight channels at once), dispense 2 μL of the kinase working stocks into columns 2 through 10 of the 384-well assay plate (Fig. 2). When dispensing the kinase working stocks from the strip tubes to 384-well plate, all eight kinases are dispensed at once. The solution of

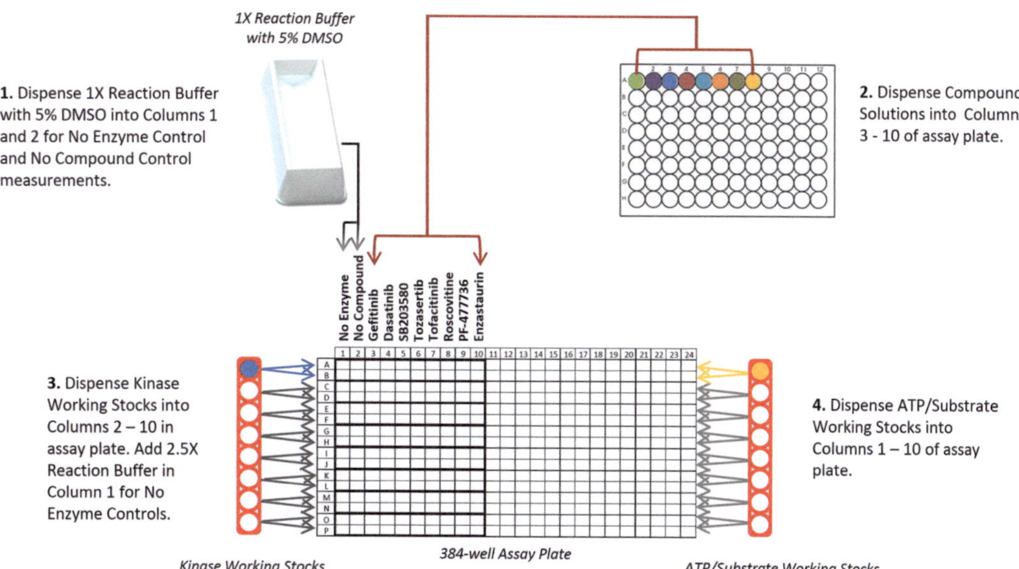

Fig. 2 Schematic representation of compound single-dose profiling setup

the first strip tube is added to row A, the solution of the second tube is added to row C, the solution of the third tube is added to row E, ending with the solution in the eighth tube added to row O (A, C, E, G, I, K, M, O).

12. Discard the tips and repeat **step 11** to dispense the kinases for the duplicate measurements by using the other wells in columns 2–10 (Fig. 2). All eight kinases are again dispensed at once to every other row starting with row B and ending with row P (B, D, F, H, J, L, N, P).
13. Cover and centrifuge the assay plate at a low setting for 15–30 s to ensure that solutions are at the bottom of the wells.
14. Place the assay plate in an incubator set at 23 °C for 10 min to allow binding of the compounds to the kinases.
15. Align the ATP/substrate working stocks in the strip, so the yellow dye is at the top relative to the 384-well plate (*see* **Note 6** and Fig. 2).
16. Dispense ATP/substrate working stocks into columns 1 through 10 of the assay plate following the same duplicate dispensing pattern setup with the kinase strip (**steps 11** and **12**).
17. Cover and centrifuge the assay plate at a low setting for 15–30 s to ensure that solutions are at the bottom of the wells.
18. Mix reaction components for 2 min on a plate shaker.
19. Place plate in incubator set at 23 °C for 60 min.
20. Detect the activity of the kinases using the ADP-Glo kinase assay as described in Subheading 3.2.

3.1.2 Inhibitor Dose–Response Profiling

The dispensing steps described in this section are shown in Fig. 3.

1. Prepare 50 μM compound solutions for both gefitinib and PF-477736 compounds. Mix 40 μL reaction buffer (2.5×), 5 μL 1 mM compound stock in DMSO, and 55 μL Nanopure water (*see* **Note 7**).
2. Use a 96-well preparative plate to prepare compound serial dilutions. Dispense 75 μL of 1× reaction buffer with 5 % DMSO into wells A2 through A12 and B2 through B12 in 96-well preparative plate (*see* **Note 8**).
3. Add 100 μL of the 50 μM gefitinib solution to well A1 of the 96-well preparative plate. Perform a 10-point 1:4 serial dilution of gefitinib across row A using 25 μL solution transfers. Mix well between transfers by pipetting up and down. Do not transfer compound into wells A11 and A12 (*see* **Note 9**).
4. Add 100 μL of the 50 μM PF-477736 solution to well B1 of the 96-well preparative plate. Perform a 10-point 1:4 serial dilution of PF-477736 across row B using 25 μL solution transfers. Mix well between transfers by pipetting up and down. Do not transfer compound solution into wells B11 and B12 (*see* **Note 9**).
5. Mix 5 μL 10 mM UltraPure ATP with 495 μL Nanopure water to make 100 μM ATP solution.
6. Thaw two substrate/cofactor strips (TK-S1) from the kit on ice.
7. Add 15 μL 100 μM ATP solution to each tube of both strips to make 2.5× ATP/substrate working stocks. Mix by pipetting up and down several times. Keep solutions on ice until use.

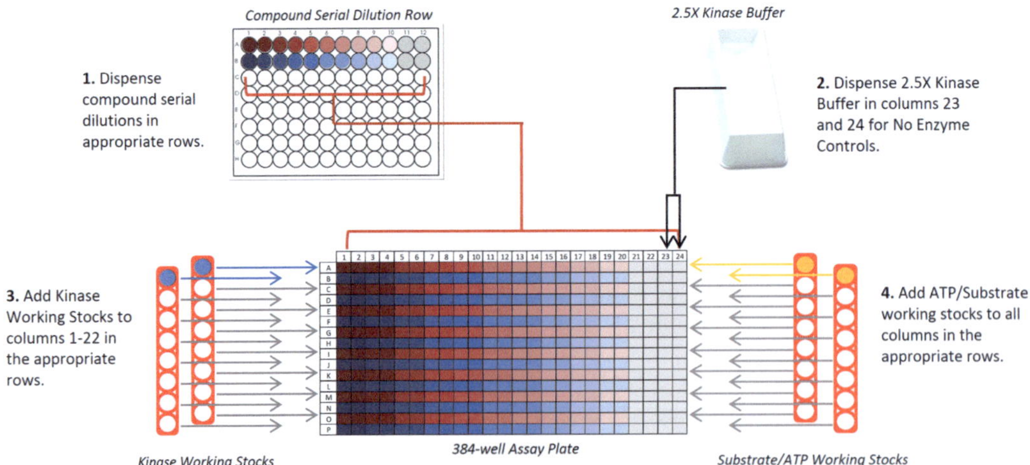

Fig. 3 Schematic representation of compound dose–response profiling setup

8. Align the 96-well preparative plate containing the serial dilutions of the compounds with the 384-well assay plate, so column 1 of the 96-well plate corresponds to columns 1 and 2 of the 384-well plate (*see* **Note 10**).
9. Use a multichannel pipette (12 channels at once) to add 1 μL of the gefitinib serial dilution from row A of the 96-well preparative plate to row A, and then to every other row of a 384-well assay plate (rows A, C, E, G, I, K, M, and O); *see* Fig. 3.
10. Use a multichannel pipette (12 channels at once) to add 1 μL of the PF-477736 serial dilution from row B of the 96-well preparative plate to the alternate rows of the 384-well assay plate (rows B, D, F, H, J, L, N, and P). Keep the plate alignment the same as outlined in **step 9**.
11. Add 2 μL of reaction buffer (2.5×) to all wells of columns 23 and 24 of the 384-well assay plate for no-enzyme control measurements.
12. Thaw both kinase strips (TK-1) from the kit on ice (*see* **Note 5**).
13. Make 2.5× kinase working stocks by adding 95 μL reaction buffer (2.5×) to each tube of both strips. Mix well by gently pipetting up and down several times. Make solutions immediately prior to use.
14. Align one of the kinase strips, so the first tube containing the blue dye lines up with row A of the 384-well assay plate (*see* **Note 6**).
15. Using a multichannel pipette (eight channels at once), add 2 μL of 2.5× kinase working stocks to wells 1 through 22 of every other row of the assay plate (row A, C, E, G, I, K, M, and O) containing the gefitinib serial dilution.
16. Align the second kinase strip, so the first tube containing the blue dye lines up with row B of the 384-well assay plate.
17. Using a multichannel pipette (eight channels at once), add 2 μL of 2.5× kinase working stocks to wells 1 through 22 of the alternate rows of the assay plate (rows B, D, F, H, J, L, N, and P) containing the PF-477736 serial dilution.
18. Cover and centrifuge the assay plate at a low setting for 15–30 s to ensure that solutions are at the bottom of the wells.
19. Place the assay plate in an incubator set at 23 °C for 10 min.
20. Align one of the substrate/cofactor strips, so the first tube containing the yellow dye lines up with row A of the 384-well assay plate.
21. Using a multichannel pipette (eight channels at once), add 2 μL of the 2.5× ATP/substrate working stocks to every other row of the assay plate (rows A, C, E, G, I, K, M, and O) (*see* **Note 11**).

22. Align the second substrate/cofactor strip, so the first tube containing the yellow dye lines up with row B of the 384-well assay plate.
23. Using a multichannel pipette (eight channels at once), add 2 μL of the 2.5× ATP/substrate working stocks to the alternate rows of the assay plate (rows B, D, F, H, J, L, N, and P).
24. Cover and centrifuge the assay plate at a low setting for 15–30 s to ensure that solutions are at the bottom of the wells.
25. Mix reaction components for 2 min on a plate shaker.
26. Place plate in incubator set at 23 °C for 60 min.
27. Detect the activity of the kinases using the ADP-Glo kinase assay as described in Subheading 3.2.

3.2 Bioluminescent Signal Detection

To detect the activity of all the kinases in the plate at once, the universal luminescence-based assay ADP-Glo is used [9]. The assay is performed in two steps; after the kinase reaction, equal volume of ADP-Glo Reagent is added to terminate the kinase reaction and to deplete the remaining ATP. After an incubation time, the kinase detection reagent is added to convert ADP to ATP and to measure the newly synthesized ATP using luciferase/luciferin reaction. The light generated is determined using a luminometer. The luminescent signal generated is proportional to the ADP concentration present and it is correlated with the amount of kinase activity.

1. Add 5 μL of ADP-Glo™ Reagent to all the wells of assay plate that have a solution.
2. Cover and centrifuge the assay plate at a low setting for 15–30 s to ensure that solutions are at the bottom of the wells.
3. Mix reaction components for 2 min on a plate shaker.
4. Place plate in incubator set at 23 °C for 40 min.
5. Add 10 μL of kinase detection Reagent to all the wells of assay plate that have a solution.
6. Cover and centrifuge the assay plate at a low setting for 15–30 s to ensure that solutions are at the bottom of the wells.
7. Mix reaction components for 2 min on a plate shaker.
8. Place plate in incubator set at 23 °C for 40 min.
9. Measure luminescence in a luminometer using an integration time of 0.5 s per well.

3.3 Profiling Data Processing

3.3.1 Single-Dose Profiling Data Processing

1. Navigate to the Kinase Selectivity Profiling System Data Analysis Worksheets on Promega Tools webpage (http://www.promega.com/resources/tools/kinase-selectivity-profiling-systems-data-analysis-worksheets/) (*see* **Note 12**).
2. Open the Single Dose Inhibition Worksheet.

a

Single-Dose Profiling

>60% Activity	20-60% Activity	<20% Activity

Kinase Selectivity Profiling System: TK-1	Kinase	Gefitinib	Dasatinib	SB203580	Tozasertib	Tofactinib	Roscovitine	PF-477736	Enzastaurin
	EGFR	1	61	96	106	84	98	39	90
	HER2	18	108	80	92	102	109	116	108
	HER4	28	12	80	97	92	86	55	86
	IGF1R	103	97	93	87	89	89	93	94
	InsR	97	86	88	73	78	74	81	82
	KDR	86	87	79	34	82	100	5	98
	PDGFRα	62	2	84	36	86	101	1	102
	PDGFRβ	85	1	81	44	80	111	1	109

Inhibitor Concentration = 1μM

Dose-Response Profiling

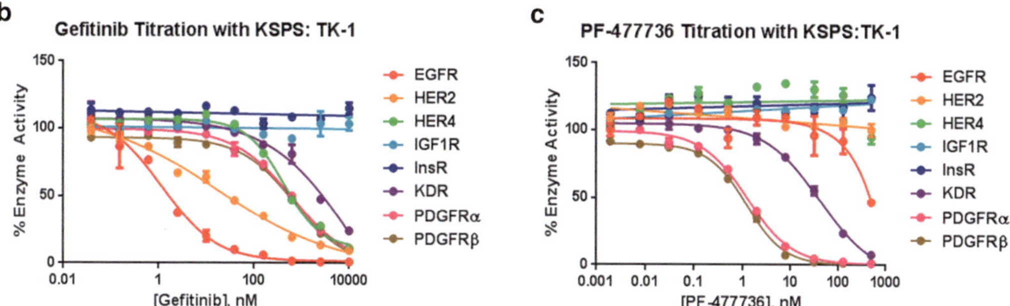

Fig. 4 (**a**) Single-dose profiling results: Percent remaining kinase activity for each kinase/compound measurement is indicated. (**b**) Gefitinib serial dilution shows selective inhibition of EGFR and HER2. (**c**) PF-477736 serial dilution shows promiscuous inhibition of kinases in the KSPS: TK-1 kit. The PF-477736 compound was identified as a CHK1 kinase inhibitor [22]

3. Copy raw luminescence data into the plate data area of the worksheet.
4. Click on the "Click to Analyze" button.
5. Choose "Columns 1–12 only (1 Strip)" option from the plate format pop-up window.
6. From the first dropdown menu select the value "8" as the "# of inhibitors tested." From the second drop-down menu select "TK-1" as the "Kinase Profiling Strip used for the assay." Click "OK."
7. In the next pop-up window, list the compounds in the order tested in the assay plate and click "OK."
8. Data is automatically processed and displayed in a color-coded table as values of the % kinase activity remaining after treatment with the individual compounds (Fig. 4a).

3.3.2 Dose–Response Profiling Data Processing

1. Navigate to the Kinase Selectivity Profiling System Data Analysis Worksheets on Promega Tools webpage (http://www.promega.com/resources/tools/kinase-selectivity-profiling-systems-data-analysis-worksheets/) (*see* **Note 12**).

2. Open the Inhibitor Dose Response Worksheet.
3. Copy raw luminescence data into the plate data area of the worksheet.
4. Click the "Click to Analyze" button.
5. Choose "Full Plate (2 Strip)" option on the Plate Format pop-up window.
6. Fill in the experimental information in the next pop-up window for strip 1 and strip 2.
   (a) Type the name of the compound tested (gefitinib for strip 1 and PF-477736 for strip 2) in the first empty field for each column under "what compound was used in the serial dilution."
   (b) Fill in the highest compound concentration (10,000 nM for both strip 1 and strip 2) in the next empty field for each column.
   (c) Fill in dilution factor (4 for both strip 1 and strip 2) in the next empty field for each column.
   (d) Select the Kinase Selectivity Profiling System used for the experiment (TK-1 for both strip 1 and strip 2) in the final drop-down menus for each column.

Click "OK" at the bottom of the pop-up window.

7. Data is automatically processed and presented as graphs of dose–response curves for each compound tested with tables indicating calculated IC50 values (Fig. 4b, c) (*see* **Note 13**).

4 Notes

1. Other compound vehicles other than DMSO may be used. Ensure that 1× buffer contains same vehicle as compounds being tested.
2. Avoid multiple freeze/thaw cycles for 10 mM UltraPure ATP stock by dispensing into smaller volumes for storage.
3. Compound solutions will contain 5 % DMSO. Make 1× buffer for controls with 5 % DMSO to match reaction conditions.
4. Single-dose plate format: Column 1 contains no-kinase controls, column 2 contains no-compound controls, and columns 3–10 contain compound measurements (*see* Fig. 2). No-kinase controls contain substrate, cofactor, and ATP in reaction buffer with 1 % DMSO. No-compound controls contain substrate, cofactor, ATP, and kinase in reaction buffer containing 1 % DMSO. Compound measurements contain substrate, cofactor, ATP, kinase, and test compound in reaction buffer

with 1 % DMSO. Duplicate measurements for each kinase are in consecutive rows (Kinase 1 duplicate measurements in rows A and B, Kinase 2 duplicate measurements in rows C and D, Kinase 3 duplicate measurements in rows E and F, Kinase 4 duplicate measurements in rows G and H, Kinase 5 duplicate measurements in rows I and J, Kinase 6 duplicate measurements in rows K and L, Kinase 7 duplicate measurements in rows M and N, and Kinase 8 duplicate measurements in rows O and P).

5. For optimal kinase activity, keep kinase strips in −80 °C freezer or on dry ice until immediately before use. Thaw kinase strips on ice when ready to dilute and use immediately. Do not reuse any remaining kinase solutions.
6. Kinase strips contain a blue dye in one tube of the eight-tube strip. Substrate/cofactor strips contain a yellow dye in one tube of the eight-tube strip. The position of the dye indicates the identity of the kit and helps with the strip orientation when setting up profiling experiments. For KSPS: TK-1 dyes are located in the first tube on the top to indicate that these strips are the first in the tyrosine kinase family kits and the kinase/substrate pair containing the dyes is EGFR/Poly(Glu_4, Tyr_1) + $MnCl_2$.
7. Make compound solutions at five times the concentration required as the highest concentration in the reactions containing the serial dilution series. Since our desired concentration range in the reaction starts at 10 μM compound, a 50 μM compound stock is used to start the serial dilution series.
8. To ensure complete dose–response curves, use a higher dilution factor in serial dilution series. We use a 1:4 dilution with the highest concentration of the compound in the reaction of 10 μM. Other values can be used to reach the desired concentration ranges.
9. We used a ten-point compound serial dilution for each compound. Column 11 contains no compound and will represent the no-compound controls. Column 12 contains no kinase and will represent the no-kinase controls.
10. Dose–response plate format: Columns 1 through 20 contain compound serial dilutions, columns 21 and 22 contain no-compound controls, and columns 23 and 24 contain no-kinase controls. Duplicate measurements are made in adjacent columns within the row of the 384-well plate. No-kinase controls contain substrate, cofactor, and ATP in reaction buffer containing 1 % DMSO. No-compound controls contain substrate, cofactor, ATP, and kinase in reaction buffer containing 1 % DMSO. Compound serial dilutions contain substrate, cofactor, ATP, kinase, and test compound in reaction buffer containing

1 % DMSO. Serial dilutions are dispensed in every other rows (rows A, C, E, G, I, K, M, and O or rows B, D, F, H, J, L, N, and P). Kinase measurements will then be located on individual rows (Kinase 1 in rows A and B, Kinase 2 in row C and D, and so on until Kinase 8 in row O and P); *see* Fig. 3.

11. When adding 2.5× ATP/substrate working stocks, start from the right side of the plate (column 24) and move towards the left. This will eliminate detectable contamination in data measurements while allowing the user fewer tip changes.
12. If plate setup matches that described in the KSPS technical manual, the KSPS Data Processing Worksheets can be used to process the generated data. The raw data output is copied into the corresponding space on the worksheet. After a few prompts to enter experimental variables, the data is processed and displayed as either a table (for single-dose profiling) or a graph (dose–response profiling). No-kinase controls are subtracted from both the no-compound controls and compound measurement values. The corrected no-compound controls are used to calculate the percentage of remaining kinase activity of the compound measurements.
13. IC50 values will not be calculated if the dose–response curve is not complete.

References

1. Hunter T (2000) Signaling-2000 and beyond. Cell 100:113–127
2. Manning G, Whyte DB, Martinez R et al (2002) The protein kinase complement of the human genome. Science 298:1912–1934
3. Cohen P (2002) Protein kinases—the major drug targets of the twenty-first century? Nat Rev Drug Discov 1:309–315
4. Widakowich C, de Castro G Jr, de Azambuja E et al (2007) Review: side effects of approved molecular targeted therapies in solid cancers. Oncologist 12:1443–1455
5. Castoldi RE, Pennella G, Saturno GS et al (2007) Assessing and managing toxicities induced by kinase inhibitors. Curr Opin Drug Discov Devel 10:53–57
6. Olaharski AJ, Gonzaludo N, Bitter H et al (2009) Identification of a Kinase Profile that Predicts Chromosome Damage Induced by Small Molecule Kinase Inhibitors. PLoS Comput Biol 5, e1000446. doi:10.1371/journal.pcbi.1000446
7. Bain J, Plater L, Elliott M et al (2007) The selectivity of protein kinase inhibitors: a further update. Biochem J 408:297–315
8. Jia Y, Quinn CM, Kwak S, Talanian RV (2008) Current in vitro kinase assay technologies: The quest for a universal format. Curr Drug Discov Technol 5:59–69
9. Zegzouti H, Zdanovskaia M, Hsiao K, Goueli SA (2009) ADP-Glo: A Bioluminescent and homogeneous ADP monitoring assay for kinases. Assay Drug Dev Technol 7:560–572
10. Krishnamurty R, Maly DJ (2007) Chemical genomic and proteomic methods for determining kinase inhibitor selectivity. Comb Chem High Throughput Screen 10:652–666
11. Miduturu CV, Deng X, Kwiatkowski N et al (2011) High-throughput kinase profiling: a more efficient approach toward the discovery of new kinase inhibitors. Chem Biol 18:868–879
12. Anastassiadis T, Deacon SW, Devarajan K et al (2011) Comprehensive assay of kinase catalytic activity reveals features of kinase inhibitor selectivity. Nat Biotechnol 29:1039–1045
13. Fabian MA, Biggs WH 3rd, Treiber DK et al (2005) A small molecule-kinase interaction map for clinical kinase inhibitors. Nat Biotechnol 23:329–336

14. Karaman MW, Herrgard S, Treiber DK et al (2008) A quantitative analysis of kinase inhibitor selectivity. Nat Biotechnol 26:127–132
15. Card M, Caldwell C, Min H et al (2009) High-Throughput Biochemical Kinase Selectivity Assays: Panel Development and Screening Applications. J Biomol Screen 14:31–42
16. Larson B, Banks P, Zegzouti H, Goueli SA (2009) A simple and robust automated kinase profiling platform using luminescent ADP accumulation technology. Assay Drug Dev Technol 7:573–584
17. Zegzouti H, Alves J, Worzella T et al (2011) Screening and profiling kinase inhibitors with a luminescent ADP detection platform. Promega Corporation Web site. http://www.promega.com/resources/pubhub/screening-and-profiling-kinase-inhibitors-with-a-luminescent-adp-detection-platform. Accessed 21 April 2015
18. Tanega C, Shen M, Mott BT et al (2009) Comparison of bioluminescent kinase assays using substrate depletion and product formation. Assay Drug Dev Technol 7:606–614
19. Davis MI, Sasaki AT, Shen M et al (2013) A Homogeneous, High-Throughput Assay for Phosphatidylinositol 5-Phosphate 4-Kinase with a Novel, Rapid Substrate Preparation. PLoS One 8, e54127. doi:10.1371/journal.pone.0054127
20. Li H, Totoritis RD, Lor LA et al (2009) Evaluation of an antibody-free ADP detection assay: ADP-Glo. Assay Drug Dev Technol 7:598–605
21. Zegzouti H, Hennek J, Alves J, Goueli S (2015) Kinase strips for routine creation of inhibitor selectivity profiles: a novel approach to targeted profiling. SLAS Conference 2015. http://www.promega.com/resources/scientific_posters/posters/kinase-strips-forcreationinhibitorselectivityprofilesnovel-approachtargetedprofilingposter/. Accessed 30 April 2015.
22. Blasina A, Hallin J, Chen E et al (2008) Breaching the DNA damage checkpoint via PF-00477736, a novel small-molecule inhibitor of checkpoint kinase 1. Mol Cancer Ther 7:2394–2404

Chapter 6

Measuring Activity of Phosphoinositide Lipid Kinases Using a Bioluminescent ADP-Detecting Assay

Andrew W. Tai and Jolanta Vidugiriene

Abstract

Phosphatidylinositol (PI) and its phosphorylated derivatives, collectively called phosphoinositides, are important second messengers involved in a variety of cellular processes, including cell proliferation, apoptosis, metabolism, and migration. These derivatives are generated by a family of kinases called phosphoinositide lipid kinases (PIKs). Due to the central role of these kinases in signaling pathways, assays for measuring their activity are often used for drug development. Lipid kinase substrates are present in unique membrane environments *in vivo* and are insoluble in aqueous solutions. Therefore the most important consideration in developing successful lipid kinase assays is the physical state of lipid kinase substrates. Here we describe the preparation of lipid substrates for two major classes of lipid kinases, phosphatidylinositol 3-kinases (PI3Ks) and phosphatidylinositol 4-kinases (PI4Ks). Using PI4Ks as an example, we also provide a detailed protocol for small-scale kinase expression and affinity purification from transiently transfected mammalian cells. For measuring lipid kinase activity we apply a universal bioluminescent ADP detection approach. The approach is compatible with diverse lipid substrates and can be used as a single integrated platform for measuring all classes of lipid and protein kinases.

Key words Lipid kinase, Phosphatidylinositol phosphate, ADP assay, Phosphatidylinositol 3-kinases, Phosphatidylinositol 4-kinases, Phosphoinositides

1 Introduction

Phosphoinositide lipid kinases (PIKs) are a family of lipid kinases that phosphorylate a hydroxyl group of the inositol ring of phosphatidylinositol (PI). Based on their ability to preferentially phosphorylate the hydroxyl group of the inositol ring at position 3, 4, or 5, PIKs have been broadly classified into three major families: phosphoinositide 3-kinases (PI3Ks), phosphoinositide 4-kinases (PI4Ks), and phosphoinositide phosphate-kinases (PIPKs) [1].

In eukaryotic cells, those kinases have preferred lipid substrates allowing for spatial and functional specificity in cell signaling and membrane remodeling [2]. These endogenous lipid substrates are

Hicham Zegzouti and Said A. Goueli (eds.), *Kinase Screening and Profiling: Methods and Protocols*, Methods in Molecular Biology, vol. 1360, DOI 10.1007/978-1-4939-3073-9_6, © Springer Science+Business Media New York 2016

found in lipid bilayers and are insoluble in aqueous solutions. To overcome solubility issues water-soluble, short acyl chain phosphoinositides, for example diC8PIP$_2$ for PI3Ks, have been used for setting up *in vitro* lipid kinase assays. However, to closer mimic *in vivo* conditions, long chain lipids are preferred substrates for *in vitro* lipid kinase assays.

For *in vitro* lipid kinase assays, synthetic liposomes are most commonly prepared by sonication [3] or extrusion (4). The lipid substrate (e.g., unphosphorylated or phosphorylated PI species for PIKs) is typically mixed with an excess of carrier lipid such as phosphatidylserine (PS) and/or phosphatidylcholine (PC). The choice of lipid carrier and the ratio of carrier:substrate can affect in vitro kinase activity, and should therefore be optimized for a given lipid kinase assay. For example, phosphatidylinositol 5-phosphate is a markedly worse substrate for PIPK IIβ in mixed PC:PI(5)P vesicles compared to PS:PI(5)P vesicles [4]. In this chapter, we present optimized protocols for using natural long chain lipids to prepare two types of functional lipid kinase substrates: liposomes for PI3K assays and lipid-detergent mixed micelles for PI4K assays.

To ensure maximum assay performance, highly purified PIKs are required. Most of these kinases are large proteins and are poorly expressed in *E. coli*. PIKs can be expressed and purified at large scales from insect cells using baculovirus expression systems [5]. For small-scale lipid kinase characterization (hundreds to thousands of assay points), expression and purification of lipid kinases from transiently transfected mammalian cells is straightforward [6, 7].

Radioactive assays based on monitoring phosphate transfer to a lipid substrate followed by lipid extraction and thin-layer chromatography (TLC) analysis are well accepted for monitoring lipid kinase activity [8]. The approach is highly sensitive and reliable but is labor intensive and not readily adaptable to miniaturization and high-throughput screening. Recently, several fluorescence and luminescence ADP detection assay formats have been developed and have gained acceptance for measuring kinase activity. The ADP detection approach is particularly attractive for lipid kinases since it can be used with diverse lipid substrates and does not require lipid extraction with organic solvents.

Here we describe detailed procedures for measuring PIK activity *in vitro* using a bioluminescent ADP detection system. We show how functional lipid substrates for two major PIKs families, PI3Ks and PI4Ks, can be prepared from long-chain natural lipids. Using FLAG-tagged PI4Ks as an example, we also provide a protocol for small-scale expression and purification of active lipid kinases from transiently transfected mammalian cells.

2 Materials

2.1 Expression and Purification of Recombinant PI4Ks

1. 293T cells (ATCC CRL-3216, or GenHunter).
2. Dulbecco's modified Eagle's medium (DMEM) with 4.5 g/L glucose, 1 mM pyruvate, 4 mM l-glutamine, supplemented with 10 % fetal bovine serum, 100 U/mL penicillin, and 100 μg/mL streptomycin.
3. Transfection reagent: Fugene HD (Promega).
4. OptiMEM-I reduced serum medium (Life Technologies).
5. Plasmids encoding N-terminal FLAG-tagged PI4Ks.
6. M2 anti-FLAG antibody (Sigma).
7. TBS: 50 mM Tris–HCl pH 7.4, 150 mM NaCl.
8. 3XFLAG peptide (Sigma F4799) dissolved in TBS to make a 5 mg/mL stock solution.
9. Protein G Dynabeads (Life Technologies) and magnetic rack.
10. Dulbecco's phosphate-buffered saline.
11. PI4K lysis buffer: 20 mM Tris–HCl pH 7.5, 100 mM NaCl, 1 % (v/v) IGEPAL CA-630 (Sigma), 10 % (v/v) glycerol, 1 mM EDTA.
12. 1 M DTT.
13. Halt protease inhibitor cocktail (Pierce) or similar.
14. Bovine serum albumin (BSA), Fraction V.

2.2 Preparation of 10× PIP_2:3DOPS Liposomes

1. PIP_2: l-α-Phosphatidylinositol-4,5-bisphosphate, ammonium salt, from porcine brain (Avanti Polar Lipids 840046X). Dissolved in chloroform/methanol/water (20:9:1).
2. DOPS: 1,2-dioleoyl-sn-glycero-3-phospho-l-serine (Avanti Polar Lipids 840035C). Dissolved in chloroform.
3. 10× Liposome dilution buffer: 250 mM HEPES pH 7.5, 5 mM EGTA.
4. Sonicators: Branson 2510 Sonifier (Danbury, CT) or Misonix S-4000 (Newton, CT).

2.3 Preparation of PI Detergent Micelles

1. PI: l-α-Phosphatidylinositol, ammonium salt, from bovine liver (Sigma P2517). Dissolved in chloroform at 10 mg/mL.
2. PI micelle buffer: 40 mM Tris–HCl pH 7.5, 20 mM $MgCl_2$, 1 mM EGTA, 0.2 % Triton X-100.
3. Sonicator: Branson S-250A Sonifier (Danbury, CT) with a cup horn or 1/8″ tapered microtip.

2.4 PI3K Assay Setup

1. 2.5× PIP_2:3DOPS liposomes: Diluted from 10× PIP_2:3DOPS liposome stock with liposome dilution buffer as described in Subheading 3.4 (*see* **Note 1**).

2. 2.5× PI3K reaction buffer: 125 mM HEPES pH 7.5, 7.5 mM $MgCl_2$, 125 mM NaCl, 0.06 mg/ml BSA.
3. 250 μM ATP: Diluted in water from 10 mM ultrapure ATP stock supplied with ADP-Glo™ Kinase Assay (Promega).
4. Recombinant Class I PI3K enzymes (*see* **Note 2**).
5. Standard solid white, multiwell plates suitable for luminescence measurements (e.g., Corning Cat. #3912, 3674).

2.5 PI4K Assay Setup

1. 2.5 mM PI micelle solution in PI micelle buffer prepared as in Subheading 3.3.
2. 2.5× PI4K buffer: 40 mM Tris–HCl pH 7.5, 20 mM $MgCl_2$, 1 mM EGTA, 0.2 % Triton X-100, 0.2 mg/mL BSA, 2 mM DTT, 1.6 mM PI, 200 μM ultrapure ATP.
3. PI4Ks: Purified as described in Subheading 3.1 or commercially available (e.g., SignalChem, EMD Millipore).

2.6 Detection of Kinase Activity Using ADP Formation

1. ADP-Glo™ Kinase Assay (Promega): Contains two reagents, the ADP-Glo™ reagent and the kinase detection reagent. Upon receiving, remove ATP and store it below −65 °C. Store the rest of the components between −30 and −10 °C.
2. Luminometer capable of reading multiwell plates. The assay data can be recorded on a variety of plate readers; although the relative light units measured will depend on the instrument.

3 Methods

3.1 Expression and Purification of Recombinant PI4Ks

For small-scale assays (hundreds to thousands of assay points), FLAG-tagged PI4Ks can be purified from transiently transfected mammalian cells. Affinity-purified kinases can be used for kinase characterization and inhibitor evaluation.

1. Plate 7.5×10^6 293T cells per 100 mm cell culture dish about 18 h before transfection. Controls should include nontransfected cells and/or cells transfected with a kinase-dead mutant PI4K construct.
2. The 293T cells should be 75–80 % confluent on the day of transfection. The following amounts are per 100 mm dish; scale up accordingly for multiple dishes.
3. Mix 5 micrograms of DNA plasmid encoding FLAG-tagged PI4K with Opti-MEM I to make a final volume of 250 μL and mix gently by flicking the tube.
4. Add 15 μL of Fugene HD to the diluted DNA and mix gently.
5. Incubate for 15 min at RT, mix gently, and then add the transfection mixture to the cells. Mix by sliding the plate back and

forth, and then side to side; do not swirl the plate in a circular motion.

6. Incubate overnight at 37 °C.
7. On the day after transfection, coat protein G-Dynabeads with anti-FLAG antibody: add 50 μL of protein G-Dynabead slurry per 100 mm dish to a microcentrifuge tube. Wash beads twice with 750 μL of PI4K lysis buffer and then resuspend in 250 μL of PI4K lysis buffer per 100 mm dish.
8. Add 5 μg of M2 anti-FLAG antibody per 100 mm dish to washed protein G-Dynabeads. Rotate tube for 20 min at RT.
9. Wash beads twice with 500 μL of PI4K lysis buffer, and finally resuspend in 100 μL of lysis buffer supplemented with 0.5 mM DTT and Halt protease inhibitor cocktail per 100 cm dish.
10. 24 h post-transfection, wash the cells twice with room temperature Dulbecco's phosphate-buffered saline. Wash gently to avoid dislodging the cell monolayer. Transfer the dish to an ice-filled tray or bucket.
11. After the last wash step, remove the PBS and add 1 mL of ice-cold PI4K lysis buffer supplemented with 0.5 mM DTT and Halt protease inhibitor cocktail per 100 cm plate and rock the dish on ice for 15 min. Collect cells by gentle pipetting and transfer to prechilled microcentrifuge tube(s) on ice.
12. Spin lysate at 14,000 × *g* for 15 min at 4 °C and transfer the supernatant to new, prechilled tube(s).
13. Add 100 μL of anti-FLAG-coated protein G-Dynabeads to each tube. Rotate tube(s) for 1 h at 4 °C.
14. Place tubes in magnetic rack and remove depleted lysate.
15. Wash beads four times with 750 μL lysis buffer with 0.5 mM DTT; with the last wash, transfer beads to a clean, prechilled microcentrifuge tube by pipetting to remove proteins nonspecifically bound to the tube wall.
16. After removing the last wash buffer, elute beads by adding 50 μL of lysis buffer with 2 mM DTT, 0.1 % BSA, and 500 μg/mL 3XFLAG peptide per 100 mm dish. Rotate for 30 min at 4 °C. Touch-spin beads briefly in microcentrifuge and then place in magnetic rack.
17. Transfer eluted FLAG-PI4K protein to a chilled clean tube. Repeat elution two more times (for a total of 150 μL per 100 mm dish).
18. Divide the purified PI4K protein into aliquots and snap-freeze in liquid nitrogen. Store at −80 °C. The lipid kinase activity is stable for at least 6 months at −80 °C.
19. Confirm purity and yield of eluted protein by SDS-PAGE and a sensitive protein stain (colloidal Coomassie Blue,fluorescent protein stain, or silver staining).

3.2 Preparation of 10× PIP_2:3DOPS Liposomes

PIP_2:3DOPS liposomes are selective substrates for class I PI3Ks (*see* **Note 3**). The liposomes are prepared at 1:3 substrate:carrier ratio for optimal performance with ADP-Glo™ detection assay.

1. Transfer 0.5 mg of PIP_2 and 1.5 mg of DOPS into a glass vial (*see* **Note 4**). Use lipid stocks dissolved in organic solvents. Dry under a stream of nitrogen in a chemical hood.
2. Rehydrate the dried lipids in 1 ml 1× liposome dilution buffer. The resulting 10× PIP_2:3DOPS suspension contains 0.5 mg/ml of PIP_2 and 1.5 mg/ml DOPS (~0.45 mM PIP_2 and 1.8 mM DOPS assuming average MW of PIP2 is 1096 g/mol and MW of DOPS is 810). The lipid suspension consisting of multilamellar vesicles will appear cloudy.
3. Sonicate in a water bath sonicator three times for 15 min at room temperature with 5-min break between cycles to produce unilamellar vesicles. The solution becomes translucent during sonication. It also can get slightly warm but it does not affect the performance of lipid substrates. To prepare larger amounts of substrates (20–50 ml) use a Misonix S-4000 with blunt-end macrotip (amplitude = 6; 10 cycles of 30 s on and 30 s off; if needed lead weights are used to hold the glassware in place).
4. Liposomes can be kept at room temperature for at least 6 h or stored at 2–10 °C for 1 week. For long-term store below –65 °C. Before using, make sure that liposomes are equilibrated to room temperature and mixed extensively by vortexing or brief resonication.

3.3 Preparation of PI Detergent Micelles

1. Dry PI (dissolved in chloroform) in a glass tube or Eppendorf brand microcentrifuge tube on ice in a fume hood under a stream of nitrogen gas, or with a centrifugal evaporator.
2. Add ice-cold PI micelle buffer to the dried PI film to a final concentration of 2.5 mM, assuming a molecular weight of 900 g/mol. Vortex the tube; the resuspended PI will appear cloudy. Sonicate the resuspended PI on ice with close monitoring until it appears translucent. For a microtip probe or cup horn sonicator, use short 3–5 s bursts at low power and allow the PI solution to chill between bursts.
3. The resulting PI detergent micelles can be stored at 4 °C for up to a month or at –20 °C for several months. The micelles should be briefly resonicated before use.

3.4 PI3K Assay Setup

This protocol describes how to set up a PI3K assay to determine the optimal enzyme concentration to be used in enzyme characterization and subsequent compound screens. The optimal amount of lipid kinase is the enzyme amount that produces luminescence within the linear range of kinase titration curve with less than 10 %

substrate conversion to product. The percent of product conversion can be determined from ATP-to-ADP conversion curves performed under the same reaction conditions (*see* **Note 5**). The protocol is provided for 25 μl reaction volumes in a 96-well plate at 25 μM ATP. Other volumes and ATP concentrations may be used (*see* **Note 6**). Representative data are shown in Fig. 1 and Table 1.

1. Prepare 2.5× PIP_2:3DOPS liposomes. Thaw 10× PIP_2:3DOPS liposome stock solution prepared in Subheading 3.2. Equilibrate to room temperature and mix extensively by vortexing. Dilute with three volumes of 1× liposome dilution buffer and vortex extensively. Keep diluted liposomes at room temperature (do not place them on ice) and vortex it again before usage.

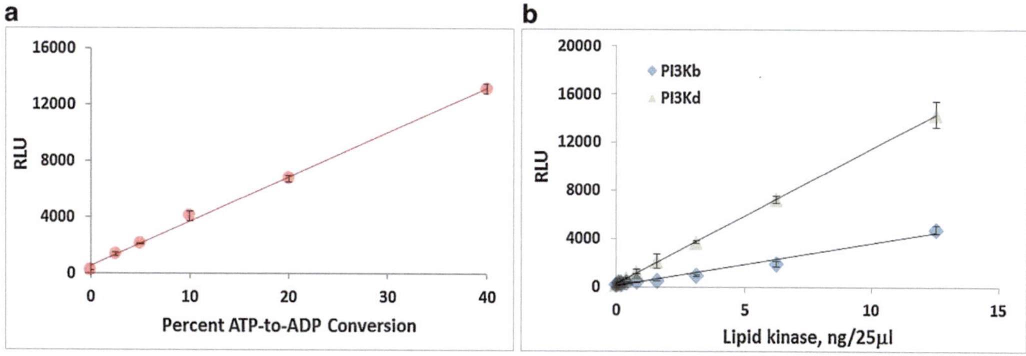

Fig. 1 Optimization of lipid kinase concentration to be used in the assay. The reaction was performed as described in PI3K assay setup protocol in 25 μl at 25 μM ATP using PIP_2:3DOPS liposomes. ATP-to-ADP conversion curve (**a**) was performed to estimate the amount of ADP produced in PI3Kβ and PI3Kδ reactions (**b**) and to determine the amount of enzyme that produces luminescence within the linear range of the kinase titration curve with less than 10 % substrate conversion to product. Each data point represents the mean +/− standard deviation of four samples

Table 1
Determining optimal enzyme concentration

ADP % (μM)	40 % (10 μM)	20 % (5 μM)	10 % (2.5 μM)	5 % (1.25 μM)	2.5 % (0.62 μM)	0 No ADP
RLUs	13052	6823	3974	2156	1353	232
Enzyme, ng/25 μl (nM)	12.5 (3 nM)	6.25 (1.5 nM)	3.12 (0.75 nM)	1.5 (0.37 nM)	0.78 (0.18 nM)	0 No enzyme
PI3Kβ-RLUs	4762	1914	945	503	419	237
PI3Kδ-RLUs	14307	7263	3714	2117	1251	211

The table show the raw relative light unit (RLU) values generated in Fig. 1.

2. To make assay buffer, combine equal volume of 2.5× PI3K reaction buffer and 2.5× PIP_2:3DOPS liposomes prepared in **step 1** (*see* **Note** 7).
3. In a preparative 96-well plate, make a twofold serial dilution of lipid kinases directly in buffer prepared in **step 2** starting at 5–10 ng/μl (*see* **Note 8**).
4. Transfer 20 μl from each well into the assay plate.
5. Start reaction by adding 5 μl of 125 μM ATP diluted in water (*see* **Note 9**).
6. Mix the plate, cover and incubate for 1 h at 23 °C (room temperature).
7. Determine kinase activity as described in Subheading 3.6.

3.5 PI4K Assay Setup

1. Thaw aliquot of purified PI4K enzyme, control(s), and ultra-pure ATP on ice. Let aliquots of the ADP-Glo reagent and reconstituted kinase detection reagent from ADP-Glo assay kit equilibrate to RT.
2. Dilute purified PI4K enzyme in PI micelle buffer with 0.1 mg/mL BSA and 2 mM DTT immediately before use (**Note 10**). The final concentration of PI4K will need to be empirically determined over a range of 1–10 ng/μL. Include blank wells with PI micelle buffer only as a no-enzyme control.
3. Make a required dilution or serial dilutions of test compounds. A positive control such as wortmannin can also be used. Transfer 1 μL of test compounds and positive controls to a solid white low-volume 384-well assay plate; we recommend performing assays in triplicate. Set up no-enzyme (background) and test compound vehicle-only (maximum signal) control wells.
4. Add 2 μL of diluted PI4K enzyme to each well. This can be done manually for a smaller number of wells or with a microplate reagent dispenser. If samples are added by manual pipetting, we recommend beginning each reaction 10 s apart and pipetting up and down three times to mix. Additional negative controls may include the use of a kinase-dead mutant PI4K or anti-FLAG immunoprecipitate from mock-transfected 293 T cells. Allow the enzyme to incubate with the test compounds for 10–20 min at RT.
5. Add 2 μL of 2.5× PI4K buffer (containing PI micelles) to each well. Negative controls may include omitting PI (to measure substrate-independent ADP generation). All volumes should be scaled proportionally for larger wells.
6. Cover the plate with a second microplate (to limit evaporation and exposure to light) and incubate at RT for 10–20 min.
7. Determine kinase activity as described in Subheading 3.6.

3.6 Detection of ADP Formation

Activity of lipid kinases is determined by measuring the amount of ADP produced by kinases using ADP-Glo™ Kinase Assay (Promega). Prepare detection reagents in advance as recommended by the manufacturer.

1. Prepare the volume of ADP-Glo™ reagent required for your experiments with 10 mM $MgCl_2$ if needed (*see* **Note 11**).
2. Stop lipid kinase reaction by adding equal volume of ADP-Glo™ reagent (for example 25 μl to 25 μl lipid kinase reaction) prepared in **step 1**.
3. Mix and incubate at 23 °C (room temperature) for 40 min.
4. Add two volumes of kinase detection reagent (for example 50 μl if lipid kinase reactions are performed in a 25 μl volume).
5. Incubate at 23 °C (room temperature) for 40 min and then read luminescence.
6. The activity of lipid kinase can be calculated by determining the increase in relative light output (RLU) in comparison to control wells (*see* **Note 12**). Inhibitor effect can be calculated by comparing relative light output in the presence and absence of inhibitor (*see* **Note 13**).

4 Notes

1. Do not dilute directly in PI3K reaction buffer since liposomes will aggregate in this buffer.
2. Store recombinant enzymes below −65 °C. At first use rapidly thaw and place on ice. Dispense any unused material into single-use aliquots and immediately snap-freeze the vials. Avoid multiple freeze-thaw cycles.
3. PIP_2 is a preferred substrate for Class I PI3Ks in vivo. However, for *in vitro* assays, either PI or PIP2 can be phosphorylated by these kinases, and depending on assay formats, both forms of lipids have been used. 10× PI:3DOPS liposomes consisting of 1 mg/ml PI and 3 mg/ml of DOPS can be prepared using the described protocol and used as a universal substrate for all members of the PI3K family.
4. Because of the inherent property of lipids to bind nonspecifically to plastics, where possible use glassware and minimize pipetting steps when working with these substrates. When preparing mixed liposomes, it is recommended to mix the component lipids solubilized in organic solvent to assure a homogenous mixture.
5. For performing ATP-to-ADP conversion curves under experimental conditions described in the protocol, make 125 μM

ATP and ADP stocks in water. Pre-mix the prepared standards at varying ratios to simulate the ATP-to-ADP concentrations at each percent of product (ADP) formation. For example, for 40 % ATP-to-ADP conversion, pre-mix 40 μl of 125 μM ADP with 60 μl of 125 μM ATP, and for 20 % ATP-to-ADP conversion, premix 20 μl ADP and 80 μl ATP. Make other dilutions accordingly. Prepare 2.5× PIP_2:3DOPS liposomes and combine them with an equal volume of 2.5× PI3K reaction buffer as described in **steps 1** and **2**. Transfer 20 μl to the assay plate. Add 5 μl of ATP + ADP standards prepared at different conversion rates. For ADP quantitation follow the protocol in Subheading 3.6. Preferably, perform the conversion curve in the same plate and at the same time as kinase reactions.

6. The assays can be performed with other ATP concentrations in an analogous manner. Other volumes may be used, provided the 1:1:2 ratio of enzyme reaction volume to ADP-Glo™ reagent volume to kinase detection reagent volumes is maintained.
7. PIP_2 liposomes should be diluted in liposome dilution buffer before combining with PI3K reaction buffer in order to minimize potential liposome aggregation at Mg^{2+} concentrations >5 mM. The final mixture can be turbid but no visible aggregates should be seen. Use it immediately as liposomes can aggregate at higher Mg^{2+} concentrations.
8. When preparing kinase working solution, dilute lipid kinases directly in assay buffer made as described in **step 2**, particularly when low enzyme concentrations are used. Lipid kinases are more stable in the presence of lipids. However, if needed, kinase reactions can be assembled using alternative protocols. For example, lipid kinases may be diluted in 10 μl of 2.5× PI3K reaction buffer and the reaction started by adding 10 μl of 2.5× PIP_2:3DOPS liposomes and 5 μl of 125 μM ATP.
9. For inhibitor testing experiments 2.5 μl of test compound or test compound vehicle can be added to the samples. After 10–20 min incubation to allow inhibitor binding to kinase, the reaction can be started by adding 2.5 μl of 250 μM ATP.
10. PI4K activity is labile. DTT or other reducing agents should be added to all buffers in which PI4Ks will be used. Store purified enzyme in single-use aliquots, thaw soon before setting up kinase assays, and discard unused enzyme.
11. To avoid liposome aggregation, the lipid kinase reactions are performed at lower Mg^{2+} concentration (<5 mM). For optimal performance of ADP-Glo™ reagent, Mg^{2+} is added to 10 mM final concentration.
12. For enzyme titration experiments, no-substrate control might be included to determine substrate-independent ATP hydro-

lysis. Substrate-independent ATP hydrolysis can indicate enzyme endogenous ATPase activity or the presence of contaminating ATP-hydrolyzing activity in the enzyme preparation. When analyzing the data, the relative light output can be compared directly if the experiments are performed under the same conditions or after background subtraction. The assay background depends on ATP concentration and purity; therefore, the use of ultrapure ATP is very important. The controls and samples must be set up under the same conditions (reaction volume, buffer, lipid substrate, and ATP concentration).

13. For inhibitor titration experiments, no-enzyme (background) and no-test compound (maximum signal) controls have to be included. To calculate the percent inhibition, subtract no-enzyme (background) control value from all data points and calculate it relative to enzyme activities in the absence of inhibitor (100 % activity).

References

1. Brown J, Auger KR (2011) Phylogenomics of phosphoinositide lipid kinases: perspectives on the evolution of second messenger signaling and drug discovery. BMC Evol Biol 11:1–14
2. Fruman DA, Meyers RE, Cantley LC (1998) Phosphoinositide kinases. Ann Rev Biochem 67:481–507
3. Vidugiriene J, Zegzouti H, Goueli SA (2009) Evaluating the utility of a bioluminescent ADP-detecting assay for lipid kinases. Assay Drug Dev Technol 96:585–597
4. Demian DJ, Clugston SL, Foster MM et al (2009) High-throughput, cell-free, liposome-based approach for assessing in vitro activity of lipid kinases. J Biomol Screen 14:838–844
5. Borawski J, Troke P, Puyang X et al (2009) Class III phosphatidylinositol 4-kinase alpha and beta are novel host factor regulators of hepatitis C virus replication. J Virol 83:10058–10074
6. Ohana RF, Hurst R, Vidugiriene J, Slater MR, Wood KV, Urh M (2011) HaloTag-based purification of functional human kinases from mammalian cells. Protein Expr Purif 76: 154–164
7. Tai AW, Bojjireddy N, Balla T (2011) A homogeneous and nonisotopic assay for phosphatidylinositol 4-kinases. Anal Biochem 417: 97–102
8. Carpenter CL, Duckworth BC, Auger KR, Cohen B, Schaffhauser BS, Cantley LC (1990) Purification and characterization of phosphoinositide 3-kinase from rat liver. J Biol Chem 265:19704–19711

Chapter 7

A High-Throughput Radiometric Kinase Assay

Krisna C. Duong-Ly and Jeffrey R. Peterson

Abstract

Aberrant kinase signaling has been implicated in a number of diseases. While kinases have become attractive drug targets, only a small fraction of human protein kinases have validated inhibitors. Screening of libraries of compounds against a kinase or kinases of interest is routinely performed during kinase inhibitor development to identify promising scaffolds for a particular target and to identify kinase targets for compounds of interest. Screening of more focused compound libraries may also be conducted in the later stages of inhibitor development to improve potency and optimize selectivity. The dot blot kinase assay is a robust, high-throughput kinase assay that can be used to screen a number of small-molecule compounds against one kinase of interest or several kinases. Here, a protocol for a dot blot kinase assay used for measuring insulin receptor kinase activity is presented. This protocol can be readily adapted for use with other protein kinases.

Key words Kinase, Kinase assay, Kinase inhibitor, High-throughput screen, Dot blot kinase assay, Insulin receptor

1 Introduction

The human kinome encodes 518 protein kinases that regulate a number of processes including cell growth, proliferation, apoptosis, metabolism, and differentiation [1]. Even though these signaling molecules are often deregulated in diseases such as cancer, few of the human kinases have approved inhibitors. One challenge in the development of any inhibitor is achieving adequate potency. Often, libraries containing diverse molecular scaffolds must be screened to determine what scaffolds result in high potency against the kinase of interest. Achieving high potency in vitro is a necessary step in developing an inhibitor that is effective at clinically achievable concentrations. Improving potency is also important for increasing the likelihood of selective activity against the target of interest. Selectivity is a major challenge in the development of kinase inhibitors since most target the highly conserved ATP-binding pocket. Due the conservation of this pocket among all

Hicham Zegzouti and Said A. Goueli (eds.), *Kinase Screening and Profiling: Methods and Protocols*, Methods in Molecular Biology, vol. 1360, DOI 10.1007/978-1-4939-3073-9_7, © Springer Science+Business Media New York 2016

kinases, compounds that bind to this region will often inhibit other kinases as well. Inhibitors that have low potency must be administered at high concentrations to be effective—at these high concentrations, off-target effects are more likely to occur, potentially resulting in dose-limiting toxicities.

One approach to circumvent these challenges is to screen large libraries of compounds against large libraries of kinases in an unbiased, target-blind manner [2]. The information extracted from such large screens can identify molecular scaffolds that potently inhibit previously unidentified targets, suggest avenues for the repurposing of existing small-molecule compounds, and provide critical information about inhibitor selectivity. Such screening efforts necessitate robust and high-throughput methods for carrying out kinase assays. These high-throughput kinase assays are often classified as either binding assays or functional assays. Binding assays measure the physical interaction between a compound and a kinase. Functional assays directly measure the change in kinase catalytic activity in the presence of an inhibitor. These approaches may not yield identical results because, for example, compounds can bind a kinase but not inhibit its catalytic activity. We recently reviewed various kinase assays that fall into these two classes [1] and both types of assays have been used in recent large-scale screens of small-molecule libraries against collections of recombinant kinases [3–6].

In this protocol, we describe a high-throughput functional assay, commonly referred to as a "dot blot kinase assay," that can be readily carried out in a biochemistry laboratory that is equipped and approved for the use of radioisotopes. This radiometric assay involves performing a kinase reaction in the presence of ATP containing a radiolabeled terminal (gamma) phosphate and a peptide substrate (Fig. 1a). The products of the reaction are spotted on phosphocellulose (P81) paper which binds the peptide substrate. The radiolabeled phosphate group that has been transferred to the bound peptide can be visualized and quantified using a phosphorimager (Fig. 1b). In addition to being used for screening, this technique can also be used to carry out dose-response measurements to determine the potency of a particular compound against a kinase of interest.

The dot blot kinase assay is considered a "gold standard" for assaying kinase functional activity as it directly measures catalytic activity using conventional ATP as the phosphate donor. Thus, unlike many binding assays, the dot blot kinase assay closely approximates the *in vivo* reaction. A key advantage of the dot blot kinase assay is the efficiency with which many kinase assays can be performed in parallel; hundreds of kinase reactions could be readily performed in a single day by a single person assuming that the kinase(s) of interest are either commercially available or can be

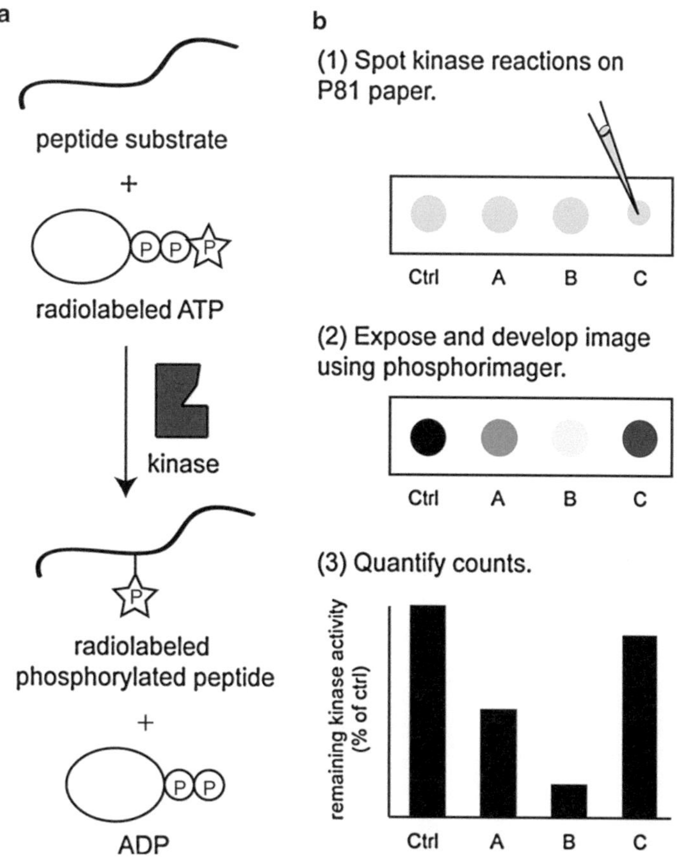

Fig. 1 (**a**) Radiometric kinase assay: In addition to conventional (unlabeled) ATP, ATP containing a radiolabeled terminal phosphate is included in the reaction mixture. Once the kinase reaction is completed, the radiolabeled phosphate is transferred to the peptide. (**b**) Dot blot capture and visualization of radiolabeled kinase reaction products: Kinase reaction products are spotted on filter paper and washed, removing excess radiolabeled ATP and facilitating peptide binding. A phosphorimager is used to visualize and quantify the counts on the filter paper for each kinase reaction

readily expressed and purified. The assay, as presented in this protocol, is best adapted to a screen in which multiple compounds are screened against a single kinase. However, this protocol can also be readily adapted for screens in which many kinases are screened against a single compound if purified kinases are available. For screens in which many kinases are screened against a library of compounds, automation may be necessary to provide adequate throughput. Commercially available screening services [3] may be more ideal in these cases.

Limitations of the assay include that the anionic P81 paper does not bind to all protein or peptide substrates equally well. The affinity of a peptide for the P81 paper depends on the number and distribution of positive charges on the peptide. To wash away unincorporated radiolabeled ATP while stabilizing the binding of substrate to the phosphocellulose matrix, sequential washes with phosphoric acid are performed. If the peptide substrate does not have sufficient positive charge in phosphoric acid, the addition of positively charged groups at the termini may be necessary during peptide synthesis [7] and must be tested empirically.

The use of a phosphorimager facilitates quantitative analysis of phosphate transfer. Autoradiography in conjunction with densitometry can be used as an alternative; however, the dynamic range and linear range of the assay in this case are much more limited [8, 9]. Several mathematical models have been developed to address this issue [10] but the implementation of these models is less convenient than phosphorimager analysis.

The specific methods outlined in this protocol for the dot blot kinase assay were used to examine small-molecule inhibition of the insulin receptor [11] but these methods can be generalized for examining inhibition of other kinase targets. Reaction conditions such as buffer components, concentration and type of substrates, as well as temperature and time may need to be optimized for kinases other than the insulin receptor kinase. Literature specific to these kinases may provide optimized reaction conditions. Another issue of critical importance that must be considered beforehand is that any kinases tested must be catalytically active and purified to a high degree (>90 % purity) to decrease the likelihood that contaminating kinases will obscure the results of the assay. One option to circumvent these difficulties is to purchase the kinase(s) of interest from a commercial source that guarantees purity and catalytic activity.

2 Materials

Prepare all buffers with ultrapure water (deionized water purified to attain a resistivity of 18 MΩ at 25 °C). All buffers may be prepared in advance and stored at the indicated temperature. The laboratory must be licensed and personnel must be trained in the proper use and disposal of radioisotopes.

2.1 Kinase Reaction

1. Compound(s) of interest: 50 μM working stock of each compound in DMSO (*see* **Notes 1** and **2**).
2. 10 mM ATP: 1 mL to be stored in 50 μL aliquots at −20 °C.
3. (γ-^{32}P)-ATP: ATP (γ-^{32}P)-3000 Ci/mmol, 5 mCi/mL EasyTide, 500 μCi (Perkin Elmer #BLU502H500UC) (*see* **Note 3**).

4. 5× IR buffer: 100 mM Hepes pH 7.4, 5 mM EGTA, 500 μM Na_3VO_4, 5 mM DTT, 10 mM $MnCl_2$, 0.1 mg/mL BSA, 50 mM $MgCl_2$. Prepare 10 mL and store 1 mL aliquots at −20 °C.
5. 1 mM peptide substrate: Poly(Glu, Tyr) 4:1 peptide, MW 5000-20000.
6. Purified insulin receptor kinase domain: Thaw one aliquot of insulin receptor kinase domain (*see* **Note 4**).

2.2 Transfer of Kinase Reaction Products to Filter Paper

1. Aluminum foil.
2. Whatman Grade P81 ion-exchange chromatography paper.
3. Pencil and ruler for marking P81 paper.
4. 0.5 % Phosphoric acid: Prepare 1 L and store at 4 °C. Use chilled solution for washes.
5. Acetone.

2.3 Visualization and Quantification

1. Phosphorimager, screen, and cassette.
2. Light box for blanking phosphorimager screen.
3. Phosphorimager quantification software.

3 Methods

3.1 Kinase Reaction

The components of a kinase reaction include buffer, ATP, a protein/peptide substrate (in the case of a protein kinase), and the kinase itself (Fig. 1a). Other components may be added to optimize the kinase reaction (*see* **Note 5**). Each assay should be performed in duplicate. We recommend using a reaction volume of 20 μL, which consists of three components:

0.2 μL Compound dissolved in DMSO or DMSO alone.

17.8 μL Buffer/substrate/kinase mixture.

2 μL ATP mixture.

For initial screening of compounds, we recommend a final compound concentration of 500 nM. Thus, 50 μM working stocks of each compound in DMSO should be prepared in advance. Here, we have chosen to use a generic peptide substrate, the poly(Glu,Tyr)4:1 peptide (*see* **Note 6**). This generic peptide is a random polymer of glutamate and tyrosine residues at a ratio of 4 glutamate residues to 1 tyrosine residues, an optimal ratio for measuring insulin receptor kinase activity [12]. The concentration of unlabeled ATP should be greater than or equal to the K_M of the kinase for ATP to ensure robust activity; here 200 μM is used. The unlabeled ATP is mixed with radioactively labeled ATP. We recommend using 0.5 μCi of (γ-^{32}P)-ATP for each reaction. At the specific activity used here, the concentration of the radioactively labeled ATP is negligible compared to the concentration of the unlabeled ATP and does not need to be computed in the final ATP concentration.

1. Preheat a 30 °C heat block with sufficient space for the number of kinase assays to be performed.
2. Use one microcentrifuge tube per compound to be tested. To each, add 0.2 μL of the appropriate 50 μM compound stock. Also, prepare a tube for the control reaction containing 0.2 μL DMSO.
3. Calculate the volumes of buffer, substrate, and kinase necessary per 20 μL reaction. For insulin receptor kinase, we use final concentrations of substrate and kinase of 80 μM and 50 nM, respectively. Calculate the volume of water required to bring the volume of the buffer/substrate/kinase mix for each reaction to 17.8 μL. Prepare sufficient buffer/substrate/kinase mix for the number of kinase assays that will be performed (*see* **Note 7**).
4. Add 17.8 μL buffer/substrate/kinase mixture to each tube.
5. Calculate the volumes of unlabeled ATP and radioactively labeled ATP necessary for each 20 μL reaction. For insulin receptor kinase, a final ATP concentration of 200 μM is used. Calculate the volume of water required to bring the volume of the ATP mixture to 2 μL. Prepare sufficient ATP mixture for the number of kinase assays that will be performed.
6. Move the experimental setup to the designated radioactivity area of the lab. Prepare the ATP mixture and add 2 μL ATP mixture to initiate each reaction.
7. Incubate the reactions at 30 °C for 30 min.
8. Freeze products of each reaction on dry ice.
9. Store reaction products at −20 °C.

3.2 Transfer of Kinase Reaction Products to Filter Paper

Take care to transfer radioactive wastes to appropriate vessels for disposal.

1. Thaw the reaction products at room temperature.
2. Using a pencil, draw a grid on the P81 filter paper consisting of 1.5 cm squares. Make sure that there are at least as many squares as there are reactions to analyze.
3. Place a piece of aluminum foil underneath the P81 paper. Spot 3 μL of each reaction in the appropriate square.
4. Allow to air-dry. This takes about 15 min in a laminar flow hood.

5. Wash the P81 paper four times with chilled 0.5 % phosphoric acid for 5 min per wash to remove free radiolabeled ATP.
6. Wash the P81 paper with acetone for 5 min. Allow the filter paper to air-dry for about 20 min in the laminar flow hood.

3.3 Visualization and Quantification

Refer to the user manual for the phosphorimager and associated quantification software for detailed instructions.

1. Blank a phosphorimager screen on a white light source for at least 10 min.
2. Expose the dried P81 paper to the sensitive surface of the phosphorimager screen for 3 h.
3. Scan the screen using the phosphorimager. If the signal is weak, re-blank the screen and re-expose for a longer period of time. If the signal is so high that it is no longer in the linear range, re-blank the screen and re-expose for a shorter period of time.
4. Measure the intensity of the spots using the phosphorimager software. Determine the percentage inhibition by dividing the intensity of an inhibited reaction by the intensity for the DMSO control reaction.

4 Notes

1. Compounds should be solubilized in DMSO at a concentration of at least 1 mM in small aliquots. Note that DMSO stocks can take >30 min to thaw. Frozen compound stocks should be stored at −20 °C.
2. DMSO is hygroscopic, meaning that it can absorb water over time. This can interfere with compound solubility. Consider purchasing small aliquots of DMSO for use in preparing stock solutions of compounds instead of continuously using DMSO from a large stock bottle.
3. Store in a properly shielded container. Using the value for the number of Ci per μL on the day of delivery and the half-life of ^{32}P, calculate the number of Ci per μL on the day of the experiment.
4. For insulin receptor kinase, it is necessary to use a baculovirus/insect cell expression system in order to produce a catalytically active kinase domain. We perform anion-exchange chromatography followed by gel filtration chromatography in order to obtain kinase that is >90 % pure by SDS-PAGE. Details of the expression and purification of the insulin receptor kinase can be found in Duong-Ly et al. [11] and Wei et al. [13]. After purification, we store the kinase in small one-time-use aliquots at −80 °C to preserve catalytic activity.

5. Phosphatase inhibitors (sodium orthovanadate for tyrosine kinases; sodium fluoride and β-glycerol phosphate for serine/threonine kinases) are often added since the reaction components may be contaminated with small amounts of phosphatases. Salt may also be necessary for kinase solubility; however, it is best to avoid using a total salt concentration of greater than 50 mM since salts can inhibit the kinase reaction.
6. Generic peptides such as poly(Glu,Tyr) and poly(Glu,Ala,Tyr) are often used as substrates for tyrosine kinases. The user should refer to literature on the kinase of interest to determine the optimal substrate to use in this assay.
7. Determine the concentrations of substrate and kinase to be used in the assay. The concentration of substrate must be in vast excess so that it is not depleted by more than 10 % by the end of the reaction. The concentration of kinase must be determined empirically and should be the minimal concentration that yields adequate signal.

Acknowledgements

We would like to thank Dr. Jinhua Wu and Dr. Stevan Hubbard for providing insulin receptor kinase baculovirus and for advice on expression and purification of the kinase. The authors' work is supported by NIH R01 GM083025 (J.R.P.), NIH R21 AI099929 (J.R.P.) and NIH T32 CA009035-36 (K.C.D.).

References

1. Duong-Ly KC, Peterson JR (2013) The human kinome and kinase inhibition. Curr Protoc Pharmacol. Chapter 2:Unit 2.9. doi:10.1002/0471141755.ph0209s60
2. Goldstein DM, Gray NS, Zarrinkar PP (2008) High-throughput kinase profiling as a platform for drug discovery. Nat Rev Drug Discov. 7:391-397. doi:10.1038/nrd2541
3. Anastassiadis T, Deacon SW, Devarajan K et al (2011) Comprehensive assay of kinase catalytic activity reveals features of kinase inhibitor selectivity. Nat Biotechnol. 29:1039-1045. doi:10.1038/nbt.2017
4. Davis MI, Hunt JP, Herrgard S et al (2011) Comprehensive analysis of kinase inhibitor selectivity. Nat Biotechnol. 29:1046-1051. doi:10.1038/nbt.1990
5. Karaman MW, Herrgard S, Treiber DK et al (2008) A quantitative analysis of kinase inhibitor selectivity. Nat Biotechnol. 26:127-132. doi:10.1038/nbt1358
6. Fabian MA, Biggs WH, Treiber DK et al (2005) A small molecule-kinase interaction map for clinical kinase inhibitors. Nat Biotechnol. 23:329-336. doi:10.1038/nbt1068
7. Hastie CJ, McLauchlan HJ, Cohen P (2006) Assay of protein kinases using radiolabeled ATP: a protocol. Nat Protoc. 1:968-971. doi:10.1038/nprot.2006.149
8. Huebner VD, Matthews HR (1985) Autoradiography of gels containing ^{32}P. Methods Mol Biol. 2:51-53. doi:10.1385/0-89603-064-4:51
9. Zouboulis CC, Tavakkol A (1994) Storage phosphor imaging technique improves the accuracy of RNA quantitation using ^{32}P-labeled cDNA probes. Biotechniques 16:290–292, 294
10. Palfi A, Hatvani L, Gulya K (1998) A new quantitative film autoradiographic method of quantifying mRNA transcripts for in situ hybridization.

J Histochem Cytochem. 46:1141-1149. doi: 10.1177/002215549804601006

11. Anastassiadis T, Duong-Ly KC, Deacon SW et al (2013) A highly selective dual insulin receptor (IR)/insulin-like growth factor 1 receptor (IGF-1R) inhibitor derived from an extracellular signal-regulated kinase (ERK) inhibitor. J Biol Chem. 288:28068-28077. doi:10.1074/jbc.M113.505032
12. Braun S, Raymond WE, Racker E (1984) Synthetic tyrosine polymers as substrates and inhibitors of tyrosine-specific protein kinases. J Biol Chem 259:2051–2054
13. Wei L, Hubbard SR, Hendrickson WA, Ellis L (1995) Expression, characterization, and crystallization of the catalytic core of the human insulin receptor protein-tyrosine kinase domain. J Biol Chem 270:8122–8130

Chapter 8

A High-Content Assay to Screen for Modulators of EGFR Function

Christophe Antczak and Hakim Djaballah

Abstract

Cell-based assays have the potential and advantage to identify cell-permeable modulators of kinase function, and hence provide an alternative to the conventional enzymatic activity-driven discovery approaches that rely on purified recombinant kinase catalytic domains. Here, we describe a domain-based high-content biosensor approach to study endogenous EGFR activity whereby EGF-induced receptor activation, subsequent trafficking, and internalization are imaged and quantified using time-dependent granule formation in cells. This method can readily be used to search for EGFR modulators in both chemical and RNAi screening; with potential applicability to other receptor tyrosine kinases.

Key words RNAi, siRNA, Iressa, Kinase, Inhibitor, High-content assay, EGFR, Cancer, Drug discovery, INCA2000, INCA3000, Microscopy, HTS

1 Introduction

Receptor tyrosine kinases (RTKs) regulate key cellular processes including proliferation, differentiation, survival, migration, and metabolism. Aberrant activities of RTK are invariably associated with carcinogenesis. RTKs, therefore, constitute important targets for cancer therapeutics exemplified by several FDA-approved small-molecule drugs targeting EGFR function, among them, gefitinib, erlotinib, and lapatinib. All three drugs belong to the 4-anilinoquinazoline class of chemicals that share a common mechanism of action by competing for ATP binding. However, the efficacy of such drugs is hampered by primary and acquired resistance as a result of mutations in the kinase domain or in the ATP-binding site of the target EGFR [1]. RTK targeting drug discovery has been focusing primarily on the use of recombinant kinase domain-based in vitro kinase assays, which unfortunately can be associated with high rate of attrition where exceptionally potent inhibitors are found to lack cellular activity. Though there have been several reported efforts for developing cell-based EGFR

Hicham Zegzouti and Said A. Goueli (eds.), *Kinase Screening and Profiling: Methods and Protocols*, Methods in Molecular Biology, vol. 1360, DOI 10.1007/978-1-4939-3073-9_8, © Springer Science+Business Media New York 2016

activation quantification [2–6], we could not identify published outcomes as to their use in chemical screening.

We have recently adopted and optimized an image-based live cell biosensor reporter platform for assaying EGFR activation, trafficking, and internalization [7]. The assay relies on stably expressed Src Homology 2 (SH2) of growth receptor-bound protein 2 (GRB2) tagged with GFP in A549, a human non-small-cell lung cancer cell line, termed A549-EGFRB cells [8]. Specifically, upon ligand-stimulated activation by epidermal growth factor (EGF), EGFR undergoes a conformational change and tyrosine phosphorylation, allowing for high-affinity binding of SH2-GRB2 to phosphotyrosine residues on the receptor kinase domain followed by EGFR-bound GFP-GRB2-SH2 intracellular granule formation. Subsequently, these formed granules can be imaged and quantified using automated fluorescent microscopy as a surrogate for measuring endogenous EGFR function. We have further validated the assay in a pilot screen against a library of 6912 chemicals and identified EGFR function modulators: 82 inhibitors and 66 activators. Among the inhibitors were 12 well-known and characterized EGFR inhibitors present in the screened library [9].

In the next sections, we guide the readers through the experimental steps for assay setup, image acquisition, image analysis, and data interpretation.

2 Materials

2.1 Cell Culture

1. The A549-EGFRB cell line was obtained as previously described [8] and can be purchased from Sigma-Aldrich (catalog # CLL1097-1VL).
2. RPMI-1640 media (Sigma-Aldrich).
3. Puromycin (Sigma-Aldrich).
4. EGF (Sigma # E-9644).
5. Dimethylsulfoxide (DMSO).
6. Triton X-100.
7. Penicillin (Thermo Fisher Scientific).
8. Streptomycin (Thermo Fisher Scientific).
9. Fetal bovine serum (FBS) (PAA Laboratories).
10. Cell culture media: RPMI-1640 media supplemented with 10 % FBS, 100 U/mL penicillin, 100 μg/mL streptomycin, and 1 μg/mL puromycin.
11. Trypsin solution: 0.05 % trypsin-EDTA (Thermo Fisher Scientific).
12. Phosphate-buffered saline (PBS) (Thermo Fisher Scientific).

13. Nuclease-free water.
14. OptiMEM media (Thermo Fisher Scientific).
15. Hoechst solution: 10 μM Hoechst 33342 in 0.05 % Triton X-100 (v/v) (Thermo Fisher Scientific).
16. Dharmafect-1 transfection Reagent (Thermo Fisher Scientific).
17. Paraformaldehyde (32 %, w/v) (Electron Microscopy Sciences).
18. PFA Solution: 4 % PFA in PBS (v/v).
19. Iressa (gefitinib) (LC Laboratories # G-4408).
20. T-175 cm^2 tissue culture flask and 384-well microtiter tissue culture-treated black plate (Corning # 3985).
21. siRNA duplex targeting EGFR (siEGFR; catalog # SKI 2027) was designed and custom synthesized at the HTS Core Facility at MSKCC [7].
22. Non-targeting control siRNA duplex (siCtrl; Thermo Fisher Scientific # 4390843).
23. Anti-EGFR mouse mAb conjugated to AF647 (Santa Cruz Biotech).
24. Goat serum.

2.2 Instrumentation (See Note 1)

1. CytoMat (Thermo Fisher Scientific): Automated temperature- and humidity-controlled incubator set at 37 °C, 5 % CO_2–95 % air.
2. PP-384-M Personal Pipettor (Apricot Designs).
3. Multidrop 384 liquid dispenser (Thermo Fisher Scientific).
4. ELx405 automated plate washer (BioTek Instruments).
5. ABgene 300 plate sealer (Thermo Fisher Scientific).
6. Imaging systems: The IN Cell Analyzer 2000 (INCA2000): automated epifluorescence microscope, and the IN Cell Analyzer 3000 (INCA3000): automated laser scanning confocal microscope (GE Healthcare, *see* **Note 2**).

2.3 Image Analysis Software

The IN Cell Developer 1.7 software (GE Healthcare): Image analysis software equipped with a variety of object segmentation analysis modules for size, shape, and intensity.

3 Methods

The basic schematic of the assay is shown in Fig. 1. In response to EGF, the activated EGF receptor internalizes resulting in granule formation. In the absence of EGF, no granule formation is observed as all receptors are localized at the cell surface in these resting cells; this would constitute the baseline of inhibiting EGF function [7].

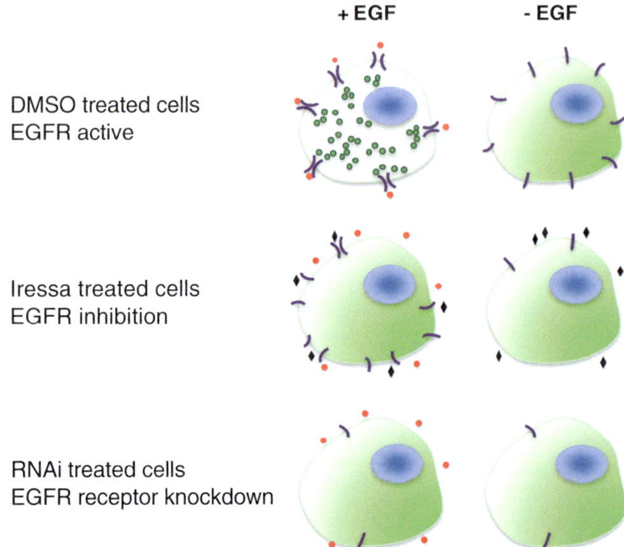

Fig. 1 EGFR biosensor assay principle: Assay schematic demonstrates the principle of the EGFR biosensor in A549-EGFRB cells in absence and in presence of EGF stimulation. GFP expression is diffuse in the cytoplasm in the absence of EGF stimulation and EGF addition leads to the recruitment and clustering of the EGFR biosensor, enabling live imaging and quantification of granule formation as a surrogate for measuring endogenous EGFR activity. EGFR biosensor activation by EGF is prevented by small-molecule EGFR inhibitor and by EGFR receptor knockdown using RNAi

3.1 Tissue Culture and Cell Maintenance

1. Maintain EGFRB cells in a 5 % CO_2-humidified incubator at 37 °C in cell culture media.
2. Passage cells every 3–4 days when reaching ~80 % cell confluency.

3.2 Assay Setup for Compounds

1. Dispense 5 μL per well of Iressa (*see* **Note 3**) from 100 μM stock in 10 % DMSO (v/v), along with 10 % DMSO (v/v) as a positive control to 384-well assay plates.
2. Dispense 2000 cells in 45 μL per well, yielding a final concentration of Iressa of 10 μM in 1 % DMSO (v/v), and incubate for 16 h at 37 °C.
3. Aspirate 35 μL of media from assay plates using the ELx405 automated plate washer.
4. For half the wells, replace with 35 μL per well of antibiotic-free media containing 500 nM EGF and for the other half supplement with 35 μL per well of antibiotic-free media.
5. Incubate for 70 min at 37 °C and perform fixing and staining as per Subheading 3.4.

3.3 Assay Setup for siRNA Duplexes

1. Dispense 2000 cells in 40 μL per well in antibiotic-free media and incubate for 24 h at 37 °C.
2. Prepare RNAi complex by mixing Dharmafect-1 transfection reagent and the siRNA duplex (siCtrl or siEGFR) in optiMEM media at a final siRNA concentration of 100 nM and 0.1 μL Dharmafect-1 reagent per well and incubate for 15 min at room temperature. This incubation time here is highly critical to achieve best knockdown.
3. Add 10 μL of the pre-prepared siRNA:lipid complex per well and incubate the cells for an additional 96 h at 37 °C.
4. Aspirate of media from assay plates using the ELx405 automated plate washer.
5. For half the wells, replace with 35 μL per well of antibiotic-free media containing 500 nM EGF and for the other half supplement with 35 μL per well of antibiotic-free media alone.
6. Incubate for 70 min at 37 °C and perform fixing and staining as per Subheading 3.4.

3.4 Cell Fixing, Nuclei Staining, and EGFR Immunostaining

1. Aspirate media from assay plates and wash cells once with 50 μL per well PBS.
2. Fix cells by dispensing 50 μL per well of PFA solution and incubate for 20 min at room temperature.
3. Wash cells once in PBS with 50 μL per well.
4. Permeabilize cells and stain nuclei by dispensing 50 μL per well of Hoechst solution and incubate for 15 min at room temperature.
5. Wash plate twice with 50 μL per well PBS.
6. Block cells with 10 % goat serum in PBS for 1.5 h at room temperature.
7. Immunostain EGFR using mouse anti-EGFR monoclonal antibody conjugated to Alexa647 and diluted 1:20 in 1 % goat serum (v/v) in PBS.
8. Incubate for 1 h at room temperature.
9. Wash plate twice with 50 μL PBS per well.
10. Dispense 50 μL of PBS per well; seal the assay plates using the ABgene plate sealer and store in the dark at 4 °C until ready for image acquisition.

3.5 Image Acquisition on the INCA2000

Images of cells in 384-well microtiter plates were acquired using the INCA2000 and imaging was performed at 20× magnification using a 20× ASAC objective (0.45NA). Images of nuclei stained with Hoechst in the blue channel were acquired using 350/50 nm excitation and 455/58 nm emission at an exposure time of 100 ms.

Images of GFP in the green channel were acquired using 490/20 nm excitation and 525/36 nm emission and GFP was imaged at an exposure time of 1.2 s [7].

3.6 Image Acquisition on the INCA3000

Images of cells in 384-well microtiter plates were acquired using the INCA3000 and imaging was performed at 40× objective magnification. Images of nuclei stained with Hoechst in the blue channel were acquired using 364 nm excitation and 450/65 nm emission in the blue channel at an exposure time of 1.5 ms. Images of GFP in the green channel were acquired using 488 nm excitation and 535/45 nm emission at an exposure time of 1.5 ms. Images of EGFR immunostaining in the red channel were acquired using 647 nm excitation and 695/55 nm emission at an exposure time of 1.5 ms [7].

3.7 Image Analysis

Images acquired by the INCA2000 were analyzed with the Developer Toolbox 1.7 software (GE Healthcare) using a custom-developed image analysis protocol. GFP granules were identified using object-based segmentation on the green channel. Hoechst-stained nuclei were identified after processing using object-based segmentation on the blue channel. Automated image analysis using a custom-developed protocol allowed us to extract granule and nuclei count, respectively, used for quantification of EGFR function and cytotoxicity [7].

3.8 Data Analysis (See Note 4)

1. Resting cells in the absence of EGF show high cell surface expression of endogenous EGFR (Fig. 2b) and upon activation with EGF, most of the expression disappears as the receptor is being internalized and forming granules (Fig. 2a).
2. Treatment of cells with non-targeting siRNA duplex (siCtrl) reveals similar results as for untreated cells (Fig. 2c, d), endogenous EGFR expression is unaffected and it is responsive to the addition of EGF. Treatment of cells with siRNA duplex targeting EGFR (siEGFR) shows near-complete knockdown of the receptor with a total absence of granules (Fig. 2f) even upon addition of EGF (Fig. 2e). These results are consistent with the granule formation being totally dependent on both the presence and activation of the EGFR.
3. To assess the robustness of this miniaturized assay, the control experiment is used consisting of several wells that contained 1 % DMSO (v/v) measuring maximum granule formation as the high control of the assay and those that contained 10 μM Iressa in 1 % DMSO (v/v) measuring complete inhibition of granule formation as the low control of the assay using the described assay workflow under Subheading 3.2.

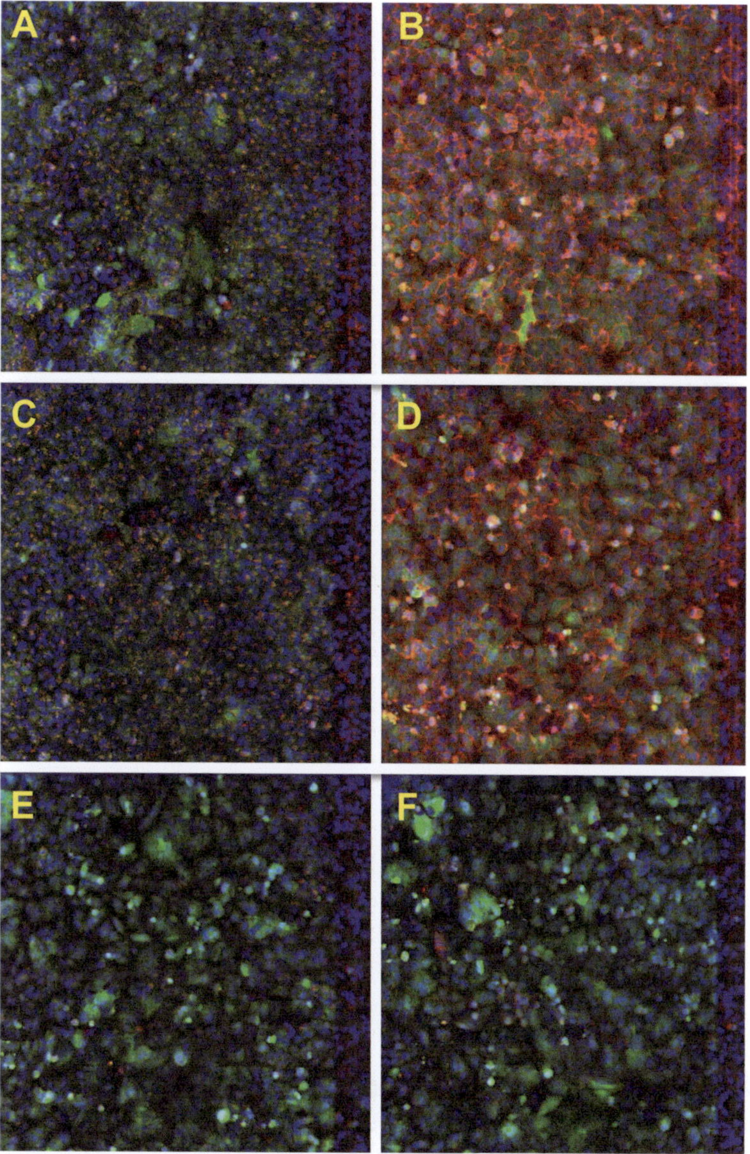

Fig. 2 Granule formation is stimulated in the presence of EGF and inhibited by RNAi. EGFR biosensor activation is induced by addition of EGF (**a**) only and reports EGFR activation; the absence of EGF shows resting cells with no apparent granules (**b**). A549-EGFRB cells treated with non-targeting siRNA duplex (siCtrl) show depletion of cell surface expression of EGFR in the presence of EGF (**c**) and not in its absence (**d**). A549-EGFRB cells treated with EGFR-targeting siRNA duplex (siEGFR) show depletion of protein at cell surface in resting cells; no added EGF (**f**); and in stimulated cells added EGF (**e**), demonstrating that granule formation is both EGF and EGFR dependent. A549-EGFRB cells imaged with the confocal microscope INCA3000 at 40× objective magnification. *Green* channel: EGF biosensor (GFP), *blue* channel: Hoechst staining of nuclei, *red* channel: immunostaining of EGFR; images are shown as an overlay of the three channels

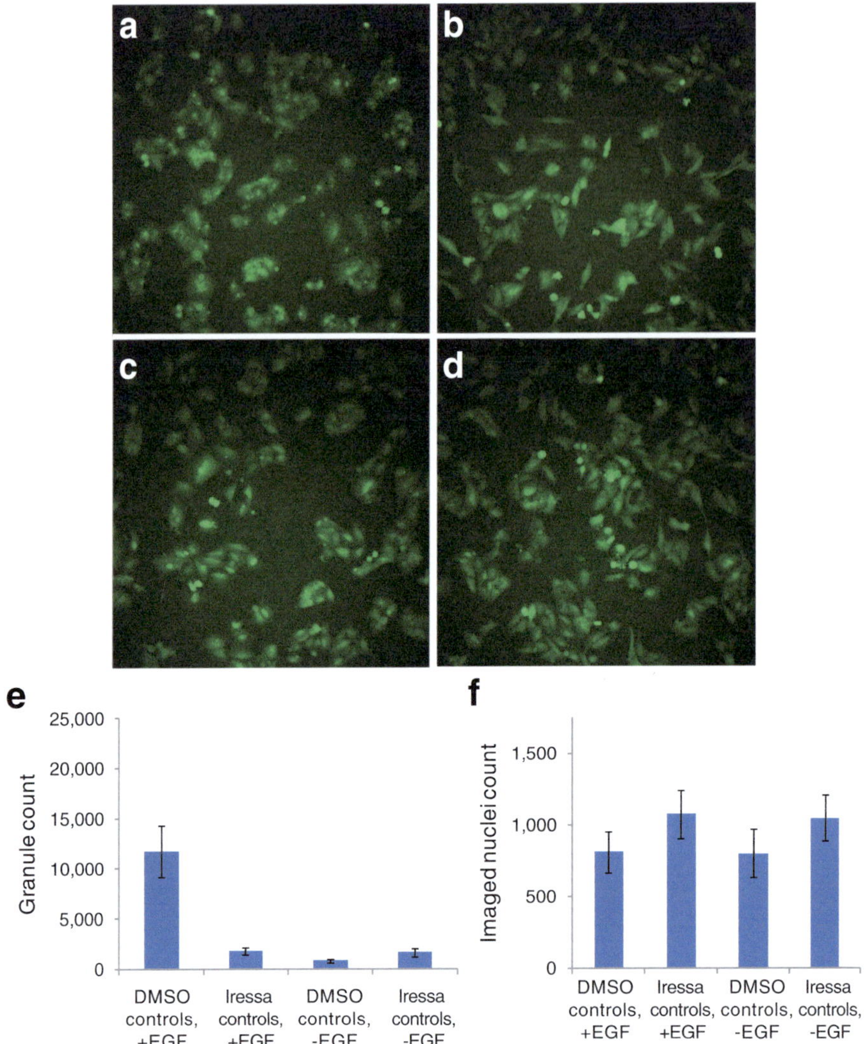

Fig. 3 EGFR granule-forming activity is inhibited by Iressa in A549-EGFRB cells. Granule formation is evident in A549-EGFRB cells, in the presence of EGF, upon treatment with 1 % DMSO (v/v) control (**a**), but totally absent in resting cells with no EGF treatment (**b**). The absence of granules is also found when cells were treated with 10 μM Iressa in 10 % DMSO (v/v) regardless of addition or not of EGF (**c**, **d**). Quantification of granule count can easily be obtained as total count per well (**e**). Quantification of nuclei count is also obtained as means to assess for cytotoxicity (**f**). A549-EGFRB cells imaged with the epifluorescence microscope INCA2000 at 20× objective magnification. *Green* channel: EGF biosensor (GFP)

4. DMSO treatment in the presence of EGF clearly shows visible granules (Fig. 3a) and a range of 10,000 to 15,000 granule count should be obtained (Fig. 3f), whereas treatment with 10 μM Iressa, in the presence of EGF, completely abrogates granule formation (Fig. 3c) with a granule count range from 0 to 1500 counts (Fig. 3f). Untreated EGF wells, regardless of DMSO or Iressa addition, did not show any substantial amount of granule counts (Fig. 3b, d).

5. Hoechst staining of cells provides an additional measure of whether the treatment was cytotoxic to the cells. Measured nuclei counts across the four treatments show little or no toxic effects (Fig. 3f). This additional measure provides a way to bin toxic versus nontoxic compounds obtained from screening chemical libraries as an example.

4 Notes

1. The list of instruments described here can easily be substituted with alternative commercially available ones.
2. Alternative automated imaging systems, such as BD Pathway 435™ (BD Biosciences, San Jose, CA); IN Cell Analyzer 1000, IN Cell Analyzer 2200, and IN Cell Analyzer 6000 (GE Healthcare); Opera and Operetta (Perkin Elmer, Waltham, MA); and BD Pathway 855™ (BD Biosciences), can easily be used for this assay.
3. Additional EGFR modulators [7, 9] can easily be purchased from commercial sources.
4. To better understand the workings of this domain-based biosensor assay, it is important to confirm that granule formation is dependent on both addition of EGF and expression of endogenous EGFP; its expression can easily be knocked down using siEGFR [7] or its activation can easily be inhibited by Iressa [7].

Acknowledgments

Work presented here was supported by Mr. William H. Goodwin and Mrs. Alice Goodwin and the Commonwealth Foundation for Cancer Research, the Experimental Therapeutics Center for MSKCC, the William Randolph Hearst Fund in Experimental Therapeutics, the Lillian S Wells Foundation, and an NIH/NCI Cancer Center Support Grant 5 P30 CA008748–44. We thank Dr. Nancy Liu-Sullivan for her help during the write-up stages of the manuscript and with the earlier versions of the assay schematic.

References

1. Pao W, Chmielecki J (2010) Rational, biologically based treatment of EGFR-mutant non-small-cell lung cancer. Nat Rev Cancer 10:760–774
2. Ciaccio MF, Wagner JP, Chuu CP et al (2010) Systems analysis of EGF receptor signaling dynamics with microwestern arrays. Nat Methods 7:148–155
3. Olive DM (2004) Quantitative methods for the analysis of protein phosphorylation in drug development. Expert Rev Proteomics 1:327–341
4. Pennucci J, Swanson S, Kaliyaperumal A et al (2010) Multiplexed evaluation of a cell-based assay for the detection of antidrug neutralizing antibodies to panitumumab in human serum

using automated fluorescent microscopy. J Biomol Screen 15:644–652

5. Ghosh RN, Grove L, Lapets OA (2004) Quantitative cell-based high-content screening assay for the epidermal growth factor receptor-specific activation of mitogen-activated protein kinase. Assay Drug Dev Technol 2:473–481
6. Lanzerstorfer P, Borgmann D, Schutz G et al (2014) Quantification and kinetic analysis of Grb2-EGFR interaction on micro-patterned surfaces for the characterization of EGFR-modulating substances. PLoS One 9(3):e92151
7. Antczak C, Bermingham A, Calder P et al (2012) Domain-based biosensor assay to screen for epidermal growth factor receptor modulators in live cells. Assay Drug Dev Technol 10:24–36
8. Sigma-Aldrich: EGFR biosensor cell line. http://www.sigmaaldrich.com/catalog/product/sigma/cll1097. Accessed 11 Mar 2014
9. Antczak C, Mahida JP, Bhinder B et al (2012) A high-content biosensor-based screen identifies cell-permeable activators and inhibitors of EGFR function: Implications in drug discovery. J Biomol Screen 17:885–899

Chapter 9

Monitoring Protein Kinase Expression and Phosphorylation in Cell Lysates with Antibody Microarrays

Hong Zhang, Xiaoqing Shi, and Steven Pelech

Abstract

Fuelled by advances in our understanding of the human kinome and phosphoproteome and the increasing availability of pan- and phosphosite-specific antibodies, antibody microarrays have emerged as powerful tools for interrogating protein phosphorylation-mediated signaling systems in *ex vivo* studies. This economical platform permits ultra-sensitive, semiquantitative measurements of the levels of hundreds of protein kinases and their substrates along with their phosphorylation status simultaneously with minute amounts of specimens. Recent technological innovations in the design and fabrication of antibody microarrays and sample preparation have permitted further refinements of the technology to yield improvements in data quality. In this chapter, we describe a detailed protocol that we have developed for tracking the expression and phosphorylation of protein kinases and their substrates in crude cell lysate samples using a high-content antibody microarray.

Key words Antibody microarray, Phosphorylation, Phosphoproteomics, Protein kinase, Protein kinome

1 Introduction

Protein phosphorylation is now recognized as the predominant reversible posttranslational regulatory mechanism in eukaryotic cells after its discovery more than half a century ago [1–3]. It underlies the control of a multitude of integrated signaling pathways that impact almost every aspect of cellular physiology. More than two-thirds of the ~21,000 proteins encoded by the human genome have been shown experimentally to be phosphorylatable at over 200,000 phosphosites [4]. Using proprietary algorithms, our most recent bioinformatics analyses of the human phosphoproteome available through the PhosphoNET KnowledgeBase (www.phosphonet.ca) have revealed that the actual number of total human phosphosites is likely to exceed one million. The reversible phosphorylation of proteins in human cells is catalyzed

Hicham Zegzouti and Said A. Goueli (eds.), *Kinase Screening and Profiling: Methods and Protocols*, Methods in Molecular Biology, vol. 1360, DOI 10.1007/978-1-4939-3073-9_9,

by at least 536 distinct human protein kinases and 156 human protein phosphatases. The biological importance and clinical significance of protein phosphorylation are plainly evident since the dysfunction of protein kinases and protein phosphatases has been associated with over 400 human diseases including cancer, diabetes, and autoimmune and neurodegenerative diseases either as a cause or consequence of these disorders [5–7].

Despite its omnipresence in eukaryotes, phosphorylation of most proteins tends to occur at low stoichiometry. For instance, only a 5 % degree of phosphorylation stoichiometry is thought to be sufficient to recruit enough of a population of protein kinases to activate signaling pathways [8, 9]. Furthermore, protein phosphorylation is dynamically regulated and therefore transient in nature, as a result of the coordinate actions of protein kinases and protein phosphatases working within complex networks with built-in redundancies and diverse feedback mechanisms. Thus, exploring the phosphoproteome calls for the use of approaches that offer high specificity and sensitivity and broad dynamic ranges.

Mass spectrometry (MS)-based phosphoproteomics is considered to be the method of choice by many in the field for identifying novel phosphorylation sites and quantifying phosphorylation events in exploratory studies despite the high equipment and consumable costs associated with this technology. Furthermore, the requirement of sample pre-processing, including protease cleavages and phosphopeptide enrichment, also poses as a potential source of data inaccuracy and bias, as evidenced by the absence of many previously characterized phosphorylation sites in the largest datasets and by the low data reproducibility observed in replicate samples [10, 11]. The need for substantial amounts of starting sample lysate protein in milligram amounts has also curtailed broad adoption of this methodology in the studies where clinical specimens are to be analyzed. While alterations in protein phosphorylation can be assessed by MS-based approaches with cultured cells, the same cannot be said about tissue samples obtained from humans and animals. Finally, the lack of antibody reagents for following up on promising leads from the MS analyses often results in reduced confidence in these findings without independent validation. Antibody microarrays define not only potentially important biomarkers, but also the antibody reagents that can serve to further track them by immunoblotting, immunoprecipitation, immunohistochemistry, and other antibody-based methods.

Antibody microarrays have emerged as an ideal complementary technology to the MS-based proteomics by virtue of their ability to analyze potentially thousands of proteins and their phosphorylation sites in a parallel fashion in crude protein lysates from cells and tissues [12]. The high specificity of capture antibodies immobilized on the array eliminates the need for other sample prefractionation steps. The growing availability of phosphosite-specific antibodies for increasing number of proteins in recent years has enhanced the capability of antibody microarrays for tracking changes not only at the protein level but also protein phosphorylation status. This is especially advantageous when working with biopsy samples that usually consist of heterogeneous cell populations.

Antibody microarrays provide a powerful and enabling platform for biomarker discovery that is unparalleled for directed proteomics analyses in terms of sensitivity and costs. A key factor in the successful application of this approach is the availability of highly specific and potent antibodies for the desired protein or phosphosite. Unfortunately, the vast majority of commercial antibodies do not meet such requirements. Even with best antibodies, another issue that can lead to false positives and false negatives with antibody microarrays is the use of cell and tissue lysates with native proteins. So as not to interfere with antibody target protein interactions, denaturing conditions are avoided in protocols with antibody microarrays. Therefore, many of the proteins captured on antibody microarrays actually reside within complexes. With homo-multimeric complexes, some antibody epitopes may become masked and the target proteins may not be able to bind to an antibody microarray. With hetero-multimeric complexes, dye labeling is also distributed on associated proteins in addition to the target proteins. Consequently, changes in the intensity of dye signals from antibody spots with perturbations of cells isolated in cultures or from treated organisms may actually arise from alterations in protein-protein interactions rather than changes in the levels or phosphorylation of the target proteins. Such detection of changes in target protein interactions with other associating proteins can, nevertheless, also provide useful insights into their regulation.

One strategy that we have taken to minimize the occurrence of protein-protein interactions is to perform chemical cleavage of lysate proteins prior to their dye labeling and incubation with antibody microarrays. This can reduce the amount of dye signal captured on typical antibody spots on microarrays by tenfold, which demonstrates the widespread occurrence of proteins within complexes. Since arginine and lysine residues commonly flank phosphorylation sites, and these may contribute to antigen recognition with phosphosite-specific antibodies, we do not recommend the use of proteases such as trypsin that cut after these basic amino acid residues to generate small peptide fragments.

Another confounding factor is the potential for capture antibodies on microarrays to cross-react with off-target proteins. Every different type of cell or tissue specimen has the possibility of featuring high levels of one or more antibody-cross-reactive proteins, which may not be predicted *a priori*. For the aforementioned reasons, it is critical to confirm any key results from antibody microarrays by immunoblotting. This can also be problematic due to the high sensitivity of antibody microarrays compared to Western blotting. Antibody microarrays provide for target protein enrichment in a much smaller area that can be achieved with immunoblots. In our experience, we find that about a quarter to a third of proteins detected on antibody microarrays fail to yield detectable protein bands on Western blots. The use of tenfold high concentrations of antibodies and lysate proteins on immunoblots can improve visualization of the target proteins, but this can be cost prohibitive.

We present here a detailed protocol that we have developed for employing the Kinex™ KAM-850 antibody microarrays to study the alterations of protein phosphorylation incurred by various perturbations or under diseased conditions. We have used the A431 human epidermoid carcinoma cell line as an example of the antibody microarray's application. A431 cells with over-expressed epidermal growth factor receptor (EGFR) are commonly employed to explore EGFR signaling mechanisms.

2 Materials

2.1 Cell line and Perturbation

1. A431 human epidermoid carcinoma cell line (CRL™-1555) (ATCC).
2. Recombinant Epidermal Growth Factor (EGF) expressed in *E. coli* (Sigma-Aldrich).

2.2 KAM-850 Antibody Microarrays

The KAM-850 antibody microarray used in the protocols described here was developed and marketed by Kinexus Bioinformatics Corporation (Vancouver, BC, Canada, www.kinexus.ca) under the trademark of Kinex™ Antibody Microarray as a component used in its proteomics services or in a stand-alone assay kit. It features over 850 well-characterized polyclonal and monoclonal antibodies selected from more than 4600 antibodies either available commercially or developed by Kinexus based on their performance on more than 25 model systems of various species origins. Among them, over 510 pan-specific antibodies provide for the detection of 189 distinct protein kinases, 31 protein phosphatases, and 142 regulatory subunits of these enzymes and other cell signaling proteins that regulate cell proliferation, growth, stress response, and apoptosis. Approximately 340 phosphosite-specific antibodies track the non-redundant phosphorylation of 128 phosphosites in protein kinases, 4 phosphosites in protein phosphatases, and 155 phosphosites in other cell signaling proteins. In addition, antibodies for proteins encoded by some housekeeping genes such as actin and tubulin and those for nonspecific immunoglobulins from antibody host species such as rabbit, mouse, and goat are also included as controls.

2.3 Lab Equipment

1. Sonicator with microprobes (Misonix) or syringes with 26-gauge needles.
2. Airfuge® refrigerated ultracentrifuge (Beckman-Coulter) or Discovery M150SE Micro-Ultracentrifuge (Thermo Scientific), or, alternatively, benchtop microcentrifuges such as the centrifuge 5454R (Eppendorf).
3. Spectrophotometer equipped with a visible light source.
4. Laser microarray scanner, e.g., ScanArray® Express Microarray Scanner (Perkin-Elmer) or GenePix 4000B Microarray Scanner (Molecular Devices).
5. Microarray imaging software: ImaGene 9.0 (Biodiscovery) or GenePix Pro 7 (Molecular Devices).
6. Other small lab equipment: Vortexer, orbital shaker, rotator, and pipettes (Gilson).

2.4 Reagents and Supplies

Albeit the use of the reagents accompanied in the kit is strongly advised for achieving the optimal performance of KAM-850 antibody microarrays, we describe the procedures herein with the use of reagents generally accessible in most laboratories as alternatives to ensure broad applicability of the protocol.

1. Dulbecco's Modified Eagle Medium (DMEM) (Life Technologies).
2. Fetal bovine serum (FBS).
3. Phosphate-buffered saline (PBS): 137 mM NaCl, 2.7 mM KCl, 10 mM Na_2HPO_4, and 2 mM KH_2PO_4, pH 7.2.
4. Protein lysis buffer: Any commercial or homemade cell lysis buffer that contains a spectrum of phosphatase inhibitors. However, it is extremely important to make sure that the lysis buffer to be used is free of Tris, glycine, or other amine-carrying components (*see* **Notes 1** and **2**).
5. cOmplete, Mini Protease Inhibitor Tablets (Roche): Alternatively, a cocktail of various protease inhibitors including 1 mM phenylmethylsulfonylfluoride (PMSF), 3 mM benzamidine, 5 μM pepstatin A, 10 μM leupeptin.
6. Reducing agents: Dithiothreitol (DTT) or Tris(2-carboxyethyl) phosphine hydrochloride (TCEP) (*see* **Note 3**).
7. Bradford Protein Assay reagent: 20 ml 95 % ethanol, 0.04 g Coomassie Brilliant Blue G-250, 40 ml 85 % phosphoric acid, adjusted the final volume with Milli-Q water to 400 ml. Alternatively, commercial Bradford assay reagents such as Bio-Rad Protein Assay Dye Reagent Concentrate (Bio-Rad) can be used.
8. Alexa Fluor®546 succinimidyl esters (NHS esters) (A20002, Life Technologies) (*see* **Note 4**).

9. 1 M $NaHCO_3$ (*see* **Note 5**).
10. Blocking buffer: 25 mM ethanolamine in 100 mM sodium borate (pH 8.5) (*see* **Note 6**).
11. Illustra MicroSpin G-25 Columns (27-5325-01, GE).
12. PAP pen (Z377821, Sigma-Aldrich).
13. Whatman® Dual well incubation chamber and Whatman® Chip Clip Slide Holder (GE, Piscataway, NJ) or a disposable KAM-850 Incubation Chamber module (Kinexus, Vancouver, BC, Canada).
14. Array incubation buffer: PBS with 0.05 % Tween-20.
15. Humidity chamber (*see* **Note 7**).
16. Wash buffer I: PBS with 2 % Tween-20.
17. Wash buffer II: PBS with 1 % Tween-20.
18. Wash buffer III: PBS with 0.5 % Tween-20.
19. Wash buffer IV: PBS with 0.1 % Tween-20.
20. Compressed N_2 (preferred) (*see* **Note 8**).

3 Methods

3.1 Cell Culture and Treatment

Culture the cells in DMEM supplemented with 10 % fetal bovine serum in final concentration, according to the protocol recommended by ATCC. Cells are starved in serum-free medium for 16 h at 60 % confluency prior to treatment. Treatments are carried out in fresh serum-free medium with or without 100 ng/ml epidermal growth factor (EGF) for 5 min.

3.2 Protein Lysate Preparation

1. Aspirate the medium from culture dishes containing approximately 1× to 2×10^6 cells for each sample at the end of the treatment (*see* **Note 9**).
2. Wash the cells on the dish briefly twice with ice-cold PBS to remove residual medium; aspirate as much PBS as possible after the last wash (*see* **Note 10**).
3. Prepare 10 ml of select lysis buffer supplemented with one cOmplete, Mini Protease Inhibitor Tablet and DTT to 1 mM final concentration (*see* **Note 11**).
4. Transfer minimal amount of ice-cold lysis buffer required to the plates and scrape the cells in the lysis buffer off the dish (*see* **Notes 12** and **13**).
5. Collect the resulting cell suspensions into 1.5 ml microcentrifuge tubes. Keep the tubes on ice (*see* **Note 14**).
6. Sonicate the cell suspension using a microprobe sonicator four times for 10 s each with 10-s intervals on ice to rupture the cells and to shear nuclear DNA. Alternatively, passing the cell

suspension through a 26-gauge needle until the sample is no longer viscous is also acceptable (*see* **Note 15**).

7. Centrifuge the resulting lysates at 90,000 × *g* or above for 30 min at 4 °C in a micro-ultracentrifuge. Clearing the lysates at maximum speed (typically 16,000-20,000 x *g*) reachable on a benchtop microcentrifuge for 30 min at 4 °C is also acceptable if a micro-ultracentrifuge is not accessible.
8. Carefully transfer the resulting supernatant to new 1.5 ml microcentrifuge tubes labeled with "A431 Control" and "A431 EGF" without disturbing the pellets at the bottom of the tubes (*see* **Note 16**).
9. Determine protein concentration of the lysates using the standard protocol of Bradford assays [13] or following the manufacturer's instruction if a commercial Bradford assay reagent is used. Bovine serum albumin (BSA) is used as the protein standard.
10. Set aside an aliquot of 100 μg of each lysate for antibody microarray analysis and store them at −20° or below if they are not to be used immediately.

3.3 Protein Labeling

1. Label two tubes containing 50 μg of Alexa Fluor®546 NHS ester dye each as "A431 Control" and "A431 EGF," respectively.
2. Transfer 50 μg of each sample into their corresponding tube.
3. Add 2.5 μl of 1 M $NaHCO_3$ to the tube and make up the volume to 30 μl with 0.1 M $NaHCO_3$ (*see* **Note 17**).
4. Vortex for 30–60 s to completely dissolve the dye.
5. Wrap the tubes in aluminum foil and place them on a rotator for 1 h at room temperature.
6. Label two Illustra MicroSpin G-25 columns with the sample names.
7. Resuspend resins in the column by finger tapping. Loosen the cap and snap off the bottom closure.
8. Place the column into a 1.5 ml microcentrifuge tube and centrifuge for 2 min at 750 × *g*. Remove the column and discard the microcentrifuge tube with liquid.
9. Place the column into a new 1.5 ml microcentrifuge tube labeled with the sample names, and transfer the entire labeling reaction volume slowly onto the top center of the resin.
10. Centrifuge for 2 min at 750 × *g*.
11. Recover the purified labeled sample from the tubes.
12. Store the dye-labeled samples at −70 °C if probing is not to be done on the same day.

3.4 Microarray Probing

1. Remove a KAM-850 antibody microarray in the vacuum-sealed package from the –70 °C freezer and leave it on the bench unopened at room temperature for 15 min to avoid moisture condensation from building up on the array surface.
2. Take out the array from the package and correctly orient the array so that the Kinexus logo and barcode at the bottom of the slide can be read correctly. Ensure that the array barcode is facing upward in the right orientation (*see* **Note 18**).
3. Hold the array to an angle, so the antibody printed areas can be clearly seen. Mark the boundaries of the printed areas on the edges along the longer sides of the array for using a PAP pen for proper placement of the incubation chamber in the next step.
4. Block the array in a tray containing sufficient amount of the blocking buffer to cover the array on an orbital shaker at room temperature for 1 h (*see* **Note 19**).
5. At the end of the blocking step, decant the blocking buffer and wash the array in the array incubation buffer for 2 min on the shaker. Repeat this step two more times.
6. Rinse the array in Milli-Q filtered H_2O for 2 min.
7. Dry the array with N_2 flow or centrifuge dry in a swing-bucket benchtop centrifuge.
8. Lower a Whatman® Dual well incubation chamber onto the top of the array so that the antibody printing areas are located within the openings of the chamber. Slide them into a Whatman® Chip Clip Slide Holder to hold them in place. Alternatively, a disposable incubation chamber module designed specifically for the KAM-850 antibody microarrays available from Kinexus can be attached onto the array surface.
9. Dilute the entire purified Alexa 546-labeled protein samples from Subheading 3.3 in 400 μl of the array incubation buffer and load the "A431 Control" and "A431 EGF" samples into the wells next to and away from the barcode end, respectively. Swirl the assembly gently to cover the array surface with the liquid completely.
10. Place the assembly on top of a humidity chamber and incubate in the dark on the shaker for 2 h at room temperature with gentle shaking.
11. Shake off the samples at the end of incubation before removing the incubation chamber.
12. Rinse the array with the wash buffer I for 1 min on the shaker with agitation.
13. Wash the array with the wash buffer II for 5 min on the shaker with agitation.

14. Wash the array with the wash buffer III for 2 min on the shaker with agitation.
15. Wash the array with the wash buffer IV for 2 min on the shaker with agitation.
16. Rinse the array in Milli-Q-purified H_2O for 2 min.
17. Dry the array with N_2 flow or centrifuge dry in a swing-bucket benchtop centrifuge.
18. Store the probed array in an array storage container and protect it from light at 4 °C.

3.5 Image Scanning and Data Analysis

1. Insert the array with the antibody side facing upwards into a microarray laser scanner. Resize the scanning areas to include the whole probing area.
2. Scan the array using 540 nm wavelength laser or equivalent at the resolution of 10 μm with the initial laser power setting at 60 %, which can be further adjusted according to the overall intensity. Additional adjustments can also be made at the photomultiplier tube (PMT) setting.
3. Images are captured and saved in the tiff format (Fig. 1).
4. The intensity of pixels within the signal and background areas of each antibody spot defined by the built-in algorithms of ImaGene 9.0 is quantified.
5. The resulting intensity readings are merged with the accompanied Gene Array List (gal) file, and the combined information is then exported and saved as output data files.

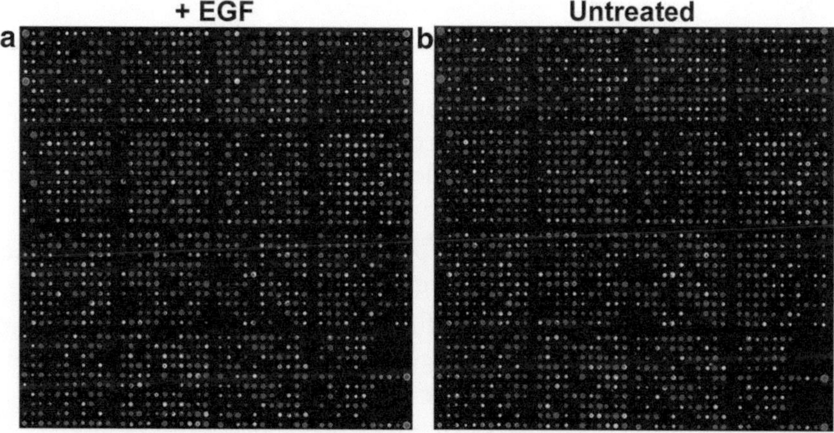

Fig. 1 An example image of two fields on a processed Kinex™ KAM-850 antibody microarray. The field (**a**) was probed with dye-labeled proteins in lysate from A341 human cervical epidermoid cells treated with 100 ng/ml of EGF for 10 min and the field (**b**) was incubated with dye-labeled proteins in lysate from untreated A431 cells. The color and size of the fluorescent spots reflected the amount of captured dye-bound protein by each of the antibodies printed on the microarray.

6. The quality of the data should be assessed based on a number of quality control parameters on the array, including spot morphologies, signal-noise ratios, background, readings from replicate spots, and the control antibodies.
7. Additional analyses including background subtraction, replicate averaging, and data normalization can be carried out to obtain a shortlist of the changes for follow-up studies through applying customized multi-parametric filtering functions such as percent change from control values, *z*-ratios, normalized signal intensity, error ranges between replicate spots, and spot quality.
8. By referring to the functional information available from relevant public databases such as Uniprot (www.uniprot.org), PhosphoNET (www.phosphonet.ca), KinaseNET (www.kinasenet.ca), and Phosphosite Plus (www.phosphosite.org), the resulting data can be interpreted in the context of known signaling pathways (Table 1).
9. To confirm the leads identified from the antibody microarray analysis, perform Western blotting analysis on the samples from the same preparation with the selected antibodies from the microarray (*see* **Notes 20** and **21**).

4 Notes

1. To preserve the phosphorylation state of the proteins, it is essential to use lysis buffers that contain a cocktail of phosphatase inhibitors for major classes of protein phosphatases such as NaF, pyrophosphate, Na_3VO_4, and β-glycerophosphate.
2. While many commercial or homemade lysis buffers may be used for preparing lysates for the KAM-850 antibody microarray analysis, they must be completely free of Tris, glycine, or other amine-carrying components, due to the amine-reactive nature of the Alexa fluorescent dye recommended in the protein labeling step.
3. TCEP is more popular due to its lower tendency to be oxidized and stronger reducing capacity than DTT.
4. The choice of fluorochrome dyes of different excitation spectra lies on the availability of the matching filters in the laser scanner used for scanning the array after probing.
5. The 1 M $NaHCO_3$ needs to be prepared fresh.
6. Ethanolamine needs to be added into 100 mM sodium borate buffer immediately prior to use due to its oxidative degradation in the presence of oxygen and CO_2.
7. The humidity chamber can be created with an empty P1000 pipette tip box. The bottom tray of the box is filled with 20 ml

Table 1

EGF-induced changes in protein phosphorylation revealed by Kinex™ KAM-850 antibody microarray analysis

Target protein	Phosphorylation site	Uniprot	Change from untreated[a]	Functional effect of phosphorylation[b]
Overall tyrosine phosphorylation				
4G10	pTyr	NA	⇧	
EGFR family of receptor-tyrosine kinases				
EGFR	T693	P00533	⇧	Inhibits phosphotransferase activity
EGFR	Y1092	P00533	⇧	Induces binding of Grb2, PLCgamma1, Ras-GAP, and PTPN6
EGFR	Y1172	P00533	⇧	Induces binding of PTPN11, Ras-GAP, and Vav2
ErbB2 (HER2)	T686	P04626	⇧	Inhibits phosphotransferase activity
ErbB2 (HER2)	Y1248	P04626	⇧	Stimulates phosphotransferase activity
ErbB3 (HER3)	Y1328	P21860	⇧	Possibly induces interaction with Arg and Shc1
MAPK pathways				
Src	Y419	P12931	⇧	Stimulates phosphotransferase activity
Raf1	S259	P04049	⇧	Inhibition of phosphotransferase activity and induces binding to 14-3-3
MEK1 (MAP2K1)	T292	Q02750	⇧	Induces binding to Raf1 and ERK2, but inhibits binding to Erk1
MEK1 (MAP2K1)	S298	Q02750	⇧	Stimulates phosphotransferase activity
MEK1 (MAP2K1)	T386	Q02750	⇧	Induces binding to ERK2
RSK1/3	T359+S363/T356+S360	Q15418	⇧	Stimulates phosphotransferase activity
RSK1/2	S380/S386	Q15418	⇧	Stimulates phosphotransferase activity
RSK1/2/3	T573/T577/T570	Q15418	⇧	Stimulates phosphotransferase activity
Elk-1	S383	P19419	⇧	Stimulates DNA binding and transcriptional activity

(continued)

Table 1
(continued)

Target protein	Phosphorylation site	Uniprot	Change from untreated[a]	Functional effect of phosphorylation[b]
ATF2	T69 + T71	P15336	⇧	Stimulates transcriptional activity
MEK3/6 (MAP2K3/6)	S218/S207	P46734	⇧	Stimulates phosphotransferase activity
p38a MAPK	T180 + Y182	Q16539	⇧	Stimulates phosphotransferase activity
JNK1/2/3	T183 + Y185	P45983	⇧	Stimulates phosphotransferase activity
Jun	T91	P05412	⇧	Stimulates phosphotransferase activity
Jun	Y170	P05412	⇧	Induces binding to Abl1
Jun	S243	P05412	⇧	Inhibits DNA binding and transcriptional activity
c-Myc	T58/S62	P01106	⇧	Stimulates transcriptional activity
Protein synthesis				
eIF2A	S52	P05198	⇧	Inhibits protein synthesis and promotes apoptosis
eIF4B	S422	P23588	⇧	Induces binding to eIF3-alpha and inhibits protein synthesis
eIF4E	S209	P06730	⇧	Increases binding to mRNA caps and formation of the EIF4F complex
eIF4G	S1107	Q04637	⇧	Unknown
eIF4G	S1232	Q04637	⇧	Unknown
4E-BP1 (PHAS1)	S65	Q13541	⇩	Inhibits binding to eIF4E and promotes protein synthesis
4E-BP1 (PHAS1)	T46	Q13541	⇩	Inhibits binding to eIF4E and promotes protein synthesis
4E-BP1 (PHAS1)	T70	Q13541	⇩	Inhibits binding to eIF4E and promotes protein synthesis
JAK-STAT signaling				
JAK1	Y1034	P23458	⇧	Stimulates phosphotransferase activity
STAT1a	Y701	P42224	⇧	Stimulates transcriptional activity
STAT2	Y690	P52630	⇧	Stimulates transcriptional activity

(continued)

Table 1 (continued)

Target protein	Phosphorylation site	Uniprot	Change from untreated[a]	Functional effect of phosphorylation[b]
PI3K pathway				
PLCgl	Y771	P19174	⇧	Unknown
PLC R (PLCg2)	Y753	P16885	⇧	Stimulates phospholipase activity and binding to Lyn
PTEN	S380+T382 +T383	P60484	⇧	Inhibits phosphatase activity
PKCd	T507	Q05655	⇧	Stimulates phosphotransferase activity
PKCq	S695	Q04759	⇧	Stimulates phosphotransferase activity and interaction with PDK1
Activation of NF-kB signaling				
IKKa	T23	O15111	⇧	Stimulates phosphotransferase activity
NFkappaB p65	S276	Q04206	⇧	Stimulates transcriptional activity
RelB	S573	Q01201	⇧	Unknown
Growth induction and proliferation				
CREB1	S133	P16220	⇧	Stimulates transcriptional activity
CREB1	S129+S133	P16220	⇧	Stimulates transcriptional activity
Cyclin B1	S147	P14635	⇧	Regulates intracellular location
Cyclin E	T395	P24864	⇧	Induces interaction with FBXW7 and SKP2, and degradation
B23 (NPM)	T234/T237	P06748	⇧	Phosphorylation possibly reduces RNA-binding activity
B23 (NPM)	T199	P06748	⇧	Phosphorylation reduces RNA-binding activity, and facilitates centrosome duplication

Dye-labeled proteins in lysate from A341 human cervical epidermoid cells treated with 100 ng/ml EGF for 5 min and dye-labeled proteins in lysate from untreated A431 cells were separately incubated with a Kinex™ KAM-850 antibody microarray, and the binding of these proteins to the individual antibodies on the microarray, were compared.

[a]Arrows indicated the directions of the changes relative to untreated control. Only changes greater than 50 % are listed.

[b]See www.phosphonet.ca for more information about the functional effects of these phosphorylations and the kinases that target these phosphosites.

of saturated NaCl solution while the porous top surface is covered with Kimwipes pre-wet with H_2O.

8. Compressed N_2 is preferred for drying the array. Alternatively, a swing-bucket centrifuge with proper slide holders can be used. Air-drying is not recommended as it may result in water marks that can lead to higher backgrounds.
9. The recovered protein lysates normally contain at least 200 μg of total protein amounts that are sufficient for antibody microarray assays. To prepare enough lysate for both antibody microarray and subsequent immunoblotting validation analyses, approximately ten times the number of the cells will be needed.
10. Serum from the culture medium must be removed completely prior to cell harvesting, or it may interfere with subsequent protein determinations.
11. The DTT and cOmplete, Mini Protease Inhibitor Tablet should be added to the lysis buffer immediately prior to use to maximize their effectiveness.
12. It is advised to minimize the initial amount of lysis buffer when recovering cells from culture dishes. Protein concentration of the lysates is ideally at 2.5 μg/μl or higher for efficient protein labeling.
13. To minimize the volume and maximize the protein concentration of lysates, lysis buffer used to recover the scrapped cells from a culture dish can be transferred to the next dish if multiple dishes of cells are to be recovered for lysate preparation.
14. Cell pellets collected should be processed in a timely fashion on ice to minimize the actions of proteases and phosphatases released during cell lysis.
15. Nuclear DNA shearing by sonication or needle passing is necessary for releasing DNA-binding proteins and for reducing the viscosity of the samples. It should not be omitted.
16. Supernatants should be separated from precipitates and frozen as quickly as possible.
17. In the case where protein concentration is below 2.5 μg/μl, samples can be concentrated by ultrafiltration using Amicon Ultra-0.5 centrifugal filter devices (Millipore), following the manufacturer's instruction.
18. Always use gloved hand to handle the array by holding it at the barcode end. Do not touch any printed surface directly.
19. Make sure that the array is always covered with buffers or lysate samples at all time after the blocking step except for **steps** 7 and **17** in Subheading 3.4.

20. Given the non-denaturing conditions applied in antibody microarray analyses for retaining the binding activity of antibodies, the tertiary and quaternary structures of protein may block access to epitopes on target proteins or phosphorylation sites recognized by the antibodies on the arrays, leading to false-negative results. Furthermore, changes in the level of an interacting protein may also be misinterpreted as alterations in the amount of the target protein or its phosphorylation state, which would result in false-positive results. In light of these possible complications, it is imperative to follow up on the leads from antibody microarray analyses with the same antibodies as those on the microarray using orthogonal approaches such as immunoblotting or immunoprecipitation. Details about the multi-immunoblotting technique we developed for validating antibody microarray results were described previously [14].

21. False negatives on antibody microarrays may include proteins that actually change in expression or phosphorylation in an experimental model system, but this is not evident from the antibody microarray analyses. In the example provided here of the Kinex™ KAM-850 antibody microarray analysis of EGF-treated--> A431 cells, changes in the phosphorylation of the MAP kinases ERK1 and ERK2 at the Thr-202/Tyr-204 and Thr-195/Tyr-197 activation sites were notably absent. This is despite the fact that two preparations of phosphosite-specific antibodies from different manufacturers were printed on the KAM-850 chips, and both antibodies were fully validated by Kinexus to work on immunoblots and revealed elevated phosphorylation in response to EGF in A431 cells. We have observed similar issues with ERK1 and ERK2 phosphorylation detection in many other types of cell lysates, including oocytes undergoing meiotic maturation from sea stars and frogs, where these kinases display the highest degrees of phosphorylation observed to date. These ERK1/ERK2 phosphosite-specific antibodies can apparently immunoprecipitate the target kinases, so it would appear that the native forms of these MAP kinases are still recognized by these antibodies in solution. This example illustrates the caution that needs to be exercised in reliance on any one methodology for proteomics analyses. Yet, despite these caveats, antibody microarrays remain the most accessible and broadly applicable tools for investigation of protein kinase signaling systems in cell and tissue specimens that are available only in limited quantities.

References

1. Burnett G, Kennedy EP (1954) The enzymatic phosphorylation of proteins. J Biol Chem 211:969–980
2. Fischer EH, Krebs EG (1955) Conversion of phosphorylase b to phosphorylase a in muscle extracts. J Biol Chem 216:121–132
3. Brognard J, Hunter T (2011) Protein kinase signaling networks in cancer. Curr Opin Genet Dev 21:4–11
4. Zhang H, Pelech S (2012) Using protein microarrays to study phosphorylation-mediated signal transduction. Semin Cell Dev Biol 23:872–882
5. Zhang J, Yang PL, Gray NS (2009) Targeting cancer with small molecule kinase inhibitors. Nat Rev Cancer 9:28–39
6. Taniguchi CM, Emanuelli B, Kahn CR (2006) Critical nodes in signalling pathways: insights into insulin action. Nat Rev Mol Cell Biol 7:85–96
7. Martin L, Latypova X, Wilson CM, Magnaudeix A, Perrin ML, Yardin C, Terro F (2013) Tau protein kinases: involvement in Alzheimer's disease. Ageing Res Rev 12:289–309
8. Mayya V, Han DK (2009) Phosphoproteomics by mass spectrometry: insights, implications, applications and limitations. Expert Rev Proteomics 6:605–618
9. Schulze WX (2010) Proteomics approaches to understand protein phosphorylation in pathway modulation. Curr Opin Plant Biol 13:280–287
10. Oyama M, Kozuka-Hata H, Tasaki S, Semba K, Hattori S, Sugano S, Inoue J, Yamamoto T (2009) Temporal perturbation of tyrosine phosphoproteome dynamics reveals the system-wide regulatory networks. Mol Cell Proteomics 8:226–231
11. Olsen JV, Vermeulen M, Santamaria A, Kumar C, Miller ML, Jensen LJ, Gnad F, Cox J, Jensen TS, Nigg EA, Brunak S, Mann M (2010) Quantitative phosphoproteomics reveals widespread full phosphorylation site occupancy during mitosis. Sci Signal 3(104):ra3
12. Zhang H, Pelech S (2012) Protein microarrays and their potential clinical applications in the era of personalized. In: Jordan B (ed) Microarrays in diagnostics and biomarker development: current and future applications. Springer, Herdelberg, Berlin, pp 55–80
13. Bradford MM (1976) A rapid and sensitive method for quantitation of microgram quantities of protein utilizing the principle of protein-dye binding. Anal Biochem 72:248–254
14. Pelech S, Sutter C, Zhang H (2003) Kinetworks™ protein kinase multiblot analysis. Methods Mol Biol 218:99–111

Chapter 10

From Enzyme to Whole Blood: Sequential Screening Procedure for Identification and Evaluation of p38 MAPK Inhibitors

Silke M. Bauer, Jakub M. Kubiak, Ulrich Rothbauer, and Stefan Laufer

Abstract

p38 mitogen-activated protein kinase (MAPK) is a pivotal enzyme in the biosynthesis of pro-inflammatory cytokines like IL-1 and TNF. Therefore, the success of anti-cytokine therapy for treatment of inflammatory processes qualified p38-MAPK as a solid target in drug research concerning chronic inflammatory diseases including infectious vascular, neurobiological, and autoimmune disorders. However, the discovery of new kinase inhibitors is limited by the need for a high biological activity combined with restricted activity to the target enzyme or pathway interaction. As a consequence, no p38 MAPK inhibitor has been introduced to the market so far, although several p38 inhibitors have proceeded into clinical trials. The development of novel inhibitor types and optimization of already known structural classes of MAPK inhibitors require appropriate testing systems reaching across these crucial parameters. As a new approach, we describe the sequential arrangement of three testing systems custom-tailored to the requirements of drug discovery programs with focus on p38 inhibition. Integrated analysis of the obtained results enables a concerted step-by-step selection of tested molecules in order to screen a compound library for the most suitable inhibitor. First, evaluation of the inhibitor's activity on the isolated p38 MAPK enzyme via an ELISA assay gives a first idea about the inhibitory potency of the molecule. Moreover, structure-activity relationships can be elucidated when comparing molecules within inhibitor series. Second, screening in living cells via a p38 substrate-specific MK2-EGFP translocation assay supplies further information about efficacy, but provides also a first notion concerning selectivity and toxicity. Third, efficacy is evaluated more specifically in vivo in LPS-stimulated human whole blood with regard to in vivo parameters, e.g., pharmacokinetic characteristics like plasma protein binding and cellular permeability. These three testing systems complement one another synergistically by providing a high overlap and predictability. Clear advantages of all presented systems are their realizability in an academic environment as well as their applicability for high-throughput screenings on a larger scale.

Key words Inhibitor screening, p38 MAPK, MK2-EGFP translocation assay, Cell-based assay, ELISA, Human whole-blood

Hicham Zegzouti and Said A. Goueli (eds.), *Kinase Screening and Profiling: Methods and Protocols*, Methods in Molecular Biology, vol. 1360, DOI 10.1007/978-1-4939-3073-9_10, © Springer Science+Business Media New York 2016

1 Introduction

The family of p38 serine/threonine mitogen-activated protein kinases (MAPK) includes four isoforms α, β, γ, and δ. These closely related homologues show more than 60 % overall sequence homology and over 90 % identity within the kinase domain, even though the four isoforms vary particularly in their tissue expression. While the α- and β-isoforms are expressed ubiquitously, the γ-isoform is preliminarily expressed in lung, pancreas, kidney, testis, and epidermis. p38 δ is restricted to tissues of skeletal muscle, heart, lung, thymus, and testis. In inflammation, the p38 MAPK cascade is considered a constitutive signaling pathway, excessively activated in autoimmune and inflammatory diseases like rheumatoid arthritis, psoriasis, Crohn's disease, and asthma, mediating a broad range of cellular responses. The MAP kinase cascade is activated by immunological, physical, chemical, or microbial stimuli, such as pro-inflammatory cytokines, LPS, hyperosmotic shock, or shear stress [1–3].

After the triggering of the aforementioned MAP kinase cascade, p38 MAPKs are activated by MAP kinase kinases (MKK) MKK3 and MKK6 via phosphorylation on conserved dual-phosphorylation sites at Thr-Gly-Tyr motifs located within the regulatory loop of the kinase. Activation of the p38 MAPKs results in phosphorylation of different substrates, such as protein kinases, e.g., MAP kinase-activated protein kinase 2 (MK2) or transcription factors, such as activating transcription factors 1, 2, and 6, among other types of substrates. With respect to the multiple intracellular transduction pathways mediated by the p38 MAPKs, the prominent connection between the p38 MAPKs and inflammatory diseases is both the regulation of gene expression at the translational level and post-transcriptional regulation of inflammatory gene expression [4]. MK2 activation by p38 MAPK phosphorylation leads to its translocation from the nucleus to the cytoplasm of the cell. Subsequent activation of MK2-related pathways or the phosphorylation of transcription factors such as ATF-2 by p38 MAPKs causes an increased expression of pro-inflammatory gene transcription resulting in the augmented concentrations of pro-inflammatory cytokines as IL-1, IL-6, and TNF [5].

To interrupt this vicious circle between the activation of p38 MAPK pathway by cytokines and the subsequent up-regulation of further cytokine production, intensive research yielded various approaches for p38 MAPK inhibition.

Herein, we report the sequential arrangement of three concerted assay systems for evaluation of inhibitory potency of p38 MAPK test compounds.

As a first approach for the preliminary selection of the most promising inhibitor candidates, the technique of ELISA allows a fast, reliable, and cost-efficient evaluation of test compounds even at a larger scale [6]. The inhibitor candidates are incipiently evaluated only with respect to their activity towards the isolated p38α MAPK enzyme. The measurement of the phosphorylation degree of activation transcription factor 2 (ATF-2) as a natural substrate of the p38 MAPKs after incubation with a candidate inhibitor provides a simple and direct readout since phosphorylation is inversely correlated with the inhibitor's potency. This allows high-throughput testing of large compound libraries which is an essential process for primary compound screening. Debarment of potential interactions between the test inhibitor and structures deriving from cell-based testing systems allows the identification of potential inhibitor candidates which can be applied as lead structures for further inhibitor series. Moreover, using only the isolated enzyme allows the investigation of structure-activity relationships between a class of structural similar molecules and the kinase in the absence of any interference factors.

Nevertheless, such an in vitro screening approach has severe limitations since it does not provide information on cellular penetrance, intracellular availability, serum depletion, toxicity, efficacy, and potency. Therefore cell-based systems have to be included in screening campaigns for novel p38 MAPK inhibitors to select promising candidates for further in vivo testing. This can be accomplished by including a versatile cell-based approach detecting the translocation of MapKap2 (MK2) in response to p38 MAPK activation/inhibition in living cells [7]. This assay is suitable for primary and secondary screening of p38 MAPK inhibitors. MK2 is a nuclear substrate of the p38 MAPK which is rapidly exported to the cytoplasm upon its p38α-mediated phosphorylation. Therefore the cytoplasmic concentration of MK2 directly correlates to p38 MAPK activity. Using a fluorescently labeled MK2 protein, the dynamic process can be visualized in microscopic time-lapse analysis of living cells [8]. Activation of the Rac/p38 signaling pathway redistributes the MK2-EGFP chimera to the cytoplasm, a fast and immediate response which is detectable within minutes to hours. Recently it has been shown that this translocation is invertible by several p38 inhibitors [9–11]. Since the MK2-EGFP translocation assay is performed in living cells it addresses highly relevant key parameters including retention upon binding to serum proteins, cellular penetrance, intracellular availability, toxicity, efficacy, and potency.

Finally, a cell-based assay for quantification of TNFα-release from human whole blood completes the screening cascade. Activated human peripheral blood monocyte (hPBMC) assays or human whole blood assays are widely used for evaluating the effectiveness of p38 MAPK inhibitors on modulation of pro-inflammatory cytokine secretion in a cell-based system. They provide a powerful tool for assessing cell activation and cytokine production in vitro. Since they imitate the natural environment, whole blood assays can be regarded as the best milieu in which to study cellular responsiveness in the presence of inhibitor candidates after LPS stimulation [12].

2 Materials

2.1 Isolated p38α MAPK ELISA

2.1.1 Labware

1. ELISA reader: EMax Precision Microplates Reader (Molecular Devices GmbH).
2. Software: SOFTmax® PRO (Molecular Devices GmbH).
3. Incubator: Titramax 1000 with incubator 1000 (Heidolph Instruments GmbH & Co. KG).
4. Vortexer: Vortex Genie 2 (Scientific Industries INC.).
5. Pipets: Eppendorf Reference® pipets, Eppendorf Research pro®, Multipette Plus® (Eppendorf AG).
6. Pipet tips: epT.I.P.S.® Standard and Combitips® plus (Eppendorf AG).
7. Assay plates: Nunc™ MaxiSorp® 96-well plates, transparent, flat-bottom (Thermo Fisher Scientific # 442404).
8. Nunc™ Microplate lid.
9. Reaction tubes: 1.5 mL, 2 mL (Eppendorf AG).
10. Falcon tubes: 15 mL, 50 mL (Sarstedt AG & Co).

2.1.2 Buffers and Solutions

1. Monoclonal Anti-phospho-ATF-2 ($pThr^{69,71}$) Peroxidase antibody (Sigma Aldrich # A6228).
2. Activating transcription factor 2 (ATF-2) (ProQinase # 0594-0000-2).
3. Active p38α MAP Kinase (Prof. Schultz, University of Tuebingen).
4. BD OptEIA™ TMB Substrate Reagent Set (BD Sciences # 555214).
5. Sulfuric acid 1 M (Merck Millipore # 109981).
6. Trizma® solution [1 M; pH 7.5]: Dissolve 2.241 g of Trizma® hydrochloride and 0.698 g of Trizma® in 20 mL of ultra-pure water.
7. Tris-buffered saline (TBS) [pH 7.5]: 50 mM Tris, and 150 mM NaCl. Mix 6.24 g Trizma® hydrochloride, 1.26 g Trizma® base and 8.77 g NaCl in 1 L of ultrapure water.
8. Adenosine 5′-triphosphate (ATP): Make 100 mM solution by dissolving 0.114 g in 2 mL of ultrapure water.
9. Bovine serum albumin (BSA) solution [10 mg/mL]: Mix 0.02 g of BSA (Sigma Aldrich # A3059) in 2 mL of ultrapure water.
10. Blocking buffer: 0.05 % TWEEN® 20 and 0.25 % BSA in TBS.
11. Antibody buffer [pH 6.5]: Blocking buffer adjusted to pH 6.5 with sodium hydroxide or hydrochloric acid respectively.
12. Coating solution: 10 μg/mL ATF-2 in TBS. Mix 55.6 μL of the activating transcription factor 2 (ATF-2) solution (648 μg/mL) with 3544.4 μL TBS in a 15 mL Falcon tube.

13. Kinase buffer (KB): 50 mM Tris 7.5, 10 mM $MgCl_2$, 10 mM β-glycerol phosphate, 100 μg/mL BSA, 1 mM DTT, 100 μM ATP, and 100 μM sodium orthovanadate.
14. p38α MAP kinase solution [241 ng/mL]: The required amount of kinase stock solution is added to the respective amount of kinase buffer to make 241 ng/mL solution and mixed carefully.
15. Anti-phospho-ATF-2 (pThr [69,71]) antibody solution [1:5000]: 2 μL of the reconstituted anti-phospho-ATF-2 (pThr[69,71]) peroxidase monoclonal antibody is added to 10 mL of antibody buffer in a 15 mL Falcon tube. The solution is mixed carefully and stored in the dark. Prepare this solution prior to use.
16. TMB substrate reagent solution: 2 mL Substrate A (H_2O_2) and 2 mL substrate B (tetramethylbenzidine). The two solutions are mixed 20 min before use in a 15 mL Falcon tube and stored in the dark.
17. Stock solutions of test compounds: A 10 mM stock solution is prepared by dissolving the respective amount of compound in pure DMSO. The stock solution was aliquoted as single-use aliquots and stored in 1.5 mL reaction tubes at –20 °C. As reference, the pyridinyl imidazole inhibitor SB 203580 is appropriate (*see* **Note 1**). Also other inhibitors may serve. A 10 mM stock solution is prepared, aliquoted, and stored as described above.

2.2 p38 Substrate-Specific MK2-EGFP Translocation Assay

2.2.1 Labware

1. Image Xpress micro XL microscope (Molecular Devices).
2. 20× Plan Apo (Nikon Instruments Europe BV).
3. Filter Set FITC-3540C-NTE-ZERO (Semrock Inc).
4. Filter Set DAPI-5060C (Semrock Inc).
5. Software: MetaXpress Custom Module Editor (Molecular Devices).
6. Incubator Cytomat 40 (Thermo Scientific).
7. Robotics for plate handling: Scara Automate.it (PAA).
8. Robotics for compound distribution: Biomek FXP (Beckman Coulter).
9. Centrifuge 5810R (Eppendorf AG).
10. Neubauer Counting Chamber (Brand GmbH).
11. T75 Cell Culture Flask (Thermo Fisher Scientific # 174952).
12. Cell culture plates: μ-clear tissue-culture micro-plates, 96 wells (Greiner Bio One).
13. Reaction tubes 1.5 mL, 2 mL (Eppendorf AG).
14. Falcon tubes 15 mL, 50 mL (Sarstedt AG & Co).

2.2.2 Chemicals and Assay Reagents

1. DMEM-High Glucose with Sodium-Pyruvate and L-Glutamine (PAA # E15-843).
2. Fetal bovine serum (FBS, PAA, #A15-151).
3. Penicillin-Streptomycin (100×, PAA, #P11-010).
4. Trypsin-EDTA (1×, PAA, #L15-004).
5. Gentamicin (10 mg/mL, PAA, P11-004).
6. G418 (Clonetech, #631309).
7. SB203580 (Merck Millipore #559400).
8. 4′,6-Diamidino-2-phenylindol (DAPI, Sigma Aldrich #D9542).
9. RNaseA.
10. Paraformaldehyde (PFA).
11. Phosphate-buffered saline (PBS) [pH 7.5]: 137 mM NaCl, 2.7 mM KCl, 10 mM Na_2HPO_4, and 1.8 mM KH_2PO_4.
12. Phosphate-buffered saline/Tween (PBST): 0.05 % Tween-20 in PBS.
13. Hanks' buffered salt solution (HBSS): 1.26 mM $CaCl_2$, 5.33 mM KCl, 0.44 mM KH_2PO_4, 0.5 mM $MgCl_2$, 0.41 mM $MgSO_4$, 138 mM NaCl, 4 mM $NaHCO_3$, 0.3 mM Na_2HPO_4, and 5.6 mM glucose.
14. Growth medium: DMEM containing 4.5 g/L glucose, 10 % FBS, 5 % PenStrep, 5 % L-glutamine, 50 μg/mL gentamicin, and 1 mg/mL G418.
15. Hyperosmotic media: Growth media with 175 mM NaCl.
16. Fixation solution (16 %): 16 % PFA in water.
17. Fixation solution (4 %): 4 % PFA in PBS.
18. DNA staining solution: 150 μg DAPI and 350 μg RNase A dissolved in 10 mL PBS (pH 7.5).
19. Stock solutions of test compounds: A 10 mM stock solution is prepared by dissolving the respective amount of compound in DMSO. Stock solutions are aliquoted into 96-well plates (Masterplate).
20. Working stock solutions: A 150 μM working stock solution is prepared by adding 1.5 μL of the stock solution to 98.5 μL of Hanks' buffered saline solution (HBSS) and 10 % DMSO.
21. Screening solution: For cellular screening, 10 μL of working solution is added to 90 μL medium yielding a final compound concentration of 15 μM. The final concentration of DMSO is 1 %. For further analysis, dilutions of lower concentrations (e.g., 1.5 μM, 150 nM, 15 nM, and 1.5 nM) can be prepared.
22. Cell line: Human U2OS-cell line stably expressing MapKap2 (MK2) N-terminal fused to the coding sequence of enhanced green fluorescent protein (EGFP): (U2OS-MK2-EGFP-clone-RA#2, Prof. Rothbauer, University of Tuebingen).

2.3 TNFα Release from Human Whole Blood Assay

2.3.1 Labware

1. ELISA reader: EMax Precision Microplate Reader (Molecular Devices GmbH).
2. Software: SOFTmax® PRO (Molecular Devices GmbH).
3. Incubator: Titramax 1000 with incubator 1000 (Heidolph Instruments GmbH & Co).
4. As ELISA plates: Nunc™ MaxiSorp® 96-well plates, transparent, flat-bottom (Thermo Fisher Scientific # 442404).
5. Nunc™ Microplate lid.
6. Reaction tubes: 1.5 mL, 2 mL.
7. Falcon tubes: 15 mL, 50 mL.
8. Deepwell plates: PP Masterblock®, 96 well, 1 mL, solid U-bottom, sterile (Greiner Bio-One # 780261).
9. Cap Mat for deepwell plate: EVA CapMat for sealing 96-well plates, round naps, alphanumeric nap coding (Greiner Bio-One # 381061).
10. As Masterplate: PP Microplate, 96 well, solid F-bottom, chimney well, alphanumeric well coding (Greiner Bio-One # 655201).
11. HERAEUS B 5060 EK-CO_2 INCUBATOR (Heraeus Laboratory Equipment GmbH). Eppendorf 5804R Centrifuge (Eppendorf AG) Pipets: Eppendorf Reference® pipets, Eppendorf Multipette Plus®, Eppendorf Research® pro multichannel pipet, Eppendorf AG, and Transferpette® electronic pipet, Brand GmbH & Cc KG.

2.3.2 Stimulation of Human Whole Blood

1. PBS [pH 7.5]: 137 mM NaCl, 2.7 mM KCl, 8.2 mM Na_2HPO_4, and 1.5 mM KH_2PO_4. Store at 4 °C for a maximum of 5 days (*see* **Note 2**).
2. DPBS-buffer: 1 g/L D-Glucose in PBS.
3. 1 % BSA/PBS buffer: 5.0 g BSA is dissolved in 500 mL of PBS buffer. Store at 4 °C for a maximum of 5 days (*see* **Note 3**).
4. Cremophor-EL/EtOH (Cr-EL/EtOH): 116.91 g Cremophor-EL (Merck # 238470) and 50 g ethanol absolute. Both components are mixed in a volumetric flask (250 mL) and the resulting solution hermetically sealed and stored in the dark for an unlimited amount of time.
5. 0.025 % DPBS-gentamicin buffer: 25.0 mg Gentamicin is diluted in 100 mL DPBS in a volumetric flask.
6. LPS solution: 10 mL of a solution containing 100 μg/mL lipopolysaccharide (LPS) (Sigma-Aldrich # L8274) in DPBS (*see* **Note 4**).
7. 1 % Cr-EL/EtOH in DPBS-gentamicin buffer: 500 μL of Cr-EL/EtOH added to 50 mL DPBS-gentamicin buffer.
8. Stock solutions of test compounds: A 10 mM stock solution is prepared by dissolving the respective amount of compound in

Cremophor-EL/EtOH. The stock solution is aliquoted in 100 μL single-use aliquots and stored in 1.5 mL reaction tubes at −20 °C. As reference, the pyridinyl imidazole inhibitor SB 203580 is appropriate. Also other inhibitors may serve (*see* **Note 5**).

9. Fresh potassium-EDTA anticoagulated whole blood from two healthy donors (donor 1 and donor 2).

2.3.3 ELISA for TNFα Quantification

1. ELISA kit: Human TNFα DuoSet (R&D Systems GmbH # DY210). Contains capture antibody: first antibody; murine anti-human TNFα antibody, detection antibody: second biotinylated caprine anti-human TNFα antibody, enzyme reagent: streptavidin-horseradish peroxidase conjugate, and recombinant human TNFα as standard. Concerning the preparation of the reconstituted reagents and their resulting concentrations, the manufacturer's instructions have to be taken into account.
2. TNFα-Diluent 2 (Beckman Coulter # IM1967): To be reconstituted with 6 mL of ultrapure water. Add water carefully into the delivered vessel. Subsequently, the vessel is covered and left for 10 min, after which it is turned upside down and left for another 10 min. Since the reconstituted solution contains serum, it will not be clear but lumps should not be visible. For further dispersion the vessel can be pivoted carefully without the generation of foam.
3. Wash buffer: PBS containing 0.05 % Tween® 20 (PBST): 2.5 mL Tween® 20 is added to 5 L of PBS buffer. For use, the solution is transferred into a wash bottle. Store at 4 °C for a maximum of 3 days.
4. BD OptEIA™ TMB Substrate Reagent Set (BD Sciences # 555214).
5. Sulfuric acid 1 M (Merck Millipore # 109981).

3 Methods

3.1 p38α MAPK ELISA Assay

3.1.1 Assay Procedure

1. Coating of the assay plate: The wells of Nunc™ MaxiSorp® assay plate are coated with 50 μL of coating solution (ATF 2). The outermost wells (columns 1 and 12 and rows A and H) remain empty (*see* Fig. 1).
2. The assay plate is covered with a lid and stored at 4 °C overnight (*see* **Notes 6** and **7**).
3. On the next day, to wash the assay plate, the coating solution is discarded and the plate is rinsed three times with double-distilled water. The plate is patted at an absorptive surface for removal of remaining water (*see* **Note 7**).

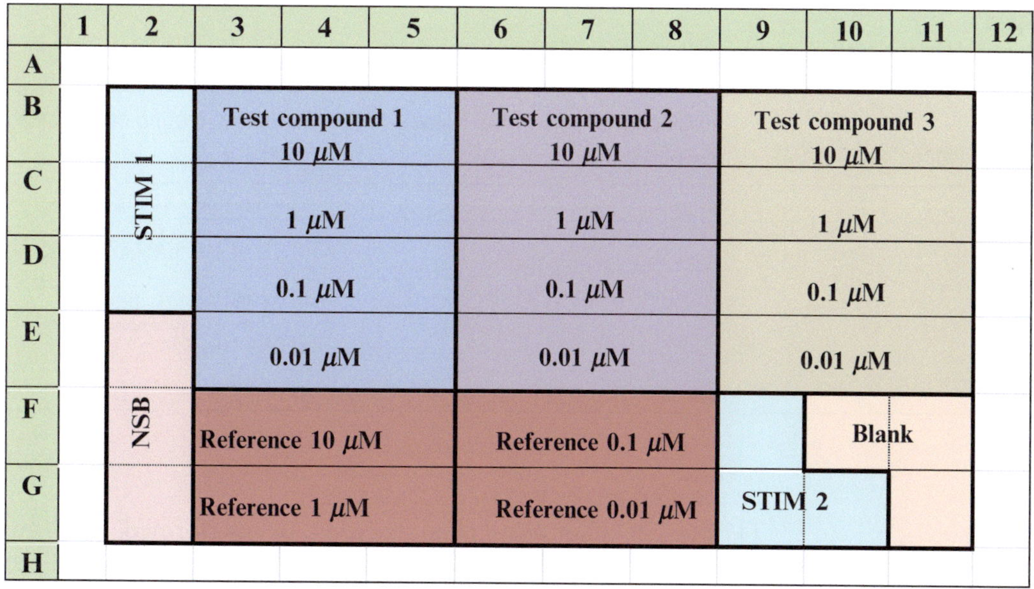

Fig. 1 Arrangement of the assay plate

4. Blocking of the unsaturated binding sites: All wells of the plate are filled with blocking buffer (*see* **Note 8**).
5. Incubate the blocking buffer for 30 min at room temperature. Meanwhile, test compounds and reference compound will be diluted.
6. Label 1.5 mL Eppendorf reaction tubes for preparation of the compound/reference dilutions. For three test compounds and the reference, 20 reaction tubes will be needed (*see* Table 1). Moreover, three more reaction tubes will be needed for STIM 1 (no-compound control 1), STIM 2 (no-compound control 2), and NSB (no-kinase control). Note that the tube support should allow cooling.
7. Prepare and hold the kinase buffer and p38α MAP kinase solution in an ice-cooled container.
8. Pipet the respective amount of p38α MAP kinase solution into the prepared Eppendorf reaction tubes for dilution of test compounds and reference (*see* Table 1).
9. Prepare a dilution row of test compounds/reference using the 10 mM stock solutions in the reaction tubes containing the kinase solution.
10. First dilution step (10 mM–100 μM): 297 μL p38α MAP kinase solution + 3 μL of the 10 mM test compound stock solution (1:100) to form a concentration of 100 μM of test compound in p38α MAP kinase solution with a remaining DMSO concentration of 1 % (*see* **Note 9**).

Table 1
Overview of the testing compound/reference dilution procedure

Final concentration (μM)	Amount of p38α MAP kinase solution (μL)	Amount of testing compound/ reference
100	297	3 μL 10 mM stock solution
10	270	30 μL of 100 μM
1	270	30 μL of 10 μM
0.1	270	30 μL of 1 μM
0.01	270	30 μL of 0.1 μM

11. Further dilutions (10–0.01 μM) are performed following Table 1: 270 μL p38α MAP kinase solution + 30 μL of the previous dilution (tenfold dilution) to obtain working concentration between 10 μM and 10 nM containing 0.1 % DMSO, at the most. Dilution procedure is carried out in an ice-cooled support. After each dilution step, the solution is mixed thoroughly.
12. Pipet 300 μL of p38α MAP kinase solution inside the reaction tubes prepared for STIM 1 and STIM 2 (*see* **step 6**).
13. Pipet 300 μL of plain KB into the Eppendorf reaction tube prepared for NSB (*see* **step 6**).
14. Preincubation step: Inhibitor dilutions, STIM 1, STIM 2, and NSB remain for 15 min in the ice-cooled support.
15. When blocking step is over, wash the assay plate as described in **step 3**. Subsequently, the plate is dried and tempered for 5 min in the incubator at 37 °C.
16. Inhibitor dilutions, STIM 1, STIM 2, and NSB are removed from the ice cooling and tempered for 3 min in the incubator at 37 °C.
17. Take the assay plate and all Eppendorf reaction tubes out of the incubator and transfer a triplicate of 50 μL of each solution to the assay plate, starting with STIM 1, NSB, test compounds 1–3, reference, and finally STIM 2. Blank wells remain empty.
18. Add 50 μL of water to the outermost wells (columns 1 and 12; rows A and H) of the plate to ensure consistent temperature during incubation.
19. The assay plate is covered with a lid and incubated for 1 h at 37 °C shaking at 150 rpm in the incubator.
20. Prepare the antibody solution of the monoclonal anti-phospho-ATF-2 (pThr ^{69,71}) peroxidase-labeled antibody.

21. Take the assay plate from the incubator and discard all solutions.
22. Washing and draining of the assay plate are performed as described in **step 3**.
23. Repeat the blocking step described in **step 4** for 15 min, and subsequently perform washing, drying, and tempering of the plate (**step 15**).
24. Apply 50 μL of the antibody solution to each well of the plate, except the outermost wells (columns 1 and 12 and rows A and H).
25. Fill the outermost wells (columns 1 and 12 and rows A and H) with water.
26. The assay plate is covered with a lid and incubated for 1 h at 37 °C shaking at 150 rpm in the incubator.
27. Take the assay plate from the incubator and discard the antibody solution.
28. Washing and draining of the assay plate were performed as described in **step 3**, followed by drying of the plate for 2 min.
29. Apply 50 μL of the TMB substrate reagent solution to each well of the plate, except the outermost wells (columns 1 and 12 and rows A and H).
30. Incubate the TMB substrate reagent solution in the dark for approximately 5 min until a deep sky blue coloring develops.
31. Stop color development by adding 25 μL of 1 M sulfuric acid to each well of the assay plate that contains TMB substrate reagent solution.
32. Read the optical densities of the assay plate in the plate reader at a wavelength of 450 nm.

3.1.2 Data Assessment

1. Subtract the mean optical density of the blank wells from every value.
2. Subtract the mean optical density of the NSB wells from every value (*see* **Note 10**).
3. The mean optical density of the STIM wells corresponds to maximum phosphorylation and is regarded as 100 % kinase activity and accordingly 0 % inhibition (*see* **Note 11**).
4. Percentage of inhibition is calculated from the mean optical density of the test compound/reference with respect to the optical density achieved for STIM according to the following equation:

$$\text{Inhibition}[\%] = 100 - \left(\frac{\text{OD}_{450}\text{Sample} - \text{OD}_{450}\text{NSB}}{\text{OD}_{450}\text{STIM} - \text{OD}_{450}\text{NSB}} \right) \times 100$$

3.2 MK2-GFP Translocation Assay

3.2.1 Assay Procedure

1. Cell culture: Thaw U2OS MK2-EGFP cells stored in cryotubes in a water bath at 37 °C. Mix defrosted cells with 5 mL growth medium and centrifuge them for 5 min at 1000 × *g*.
2. Eliminate the supernatant and resuspend cells in a 5 mL fresh growth media.
3. Cultivate cells in T75 cell culture flasks. Split the cells at 75 % confluence using trypsin to detach cells from flask. In preparation for screening experiments, detached cells have to be counted in a Neubauer counting chamber.
4. Seed the cells by disseminating ~12.500 U2OS MK2-EGFP cells in 100 μL growth medium into wells of 96-well clear tissue-culture micro-plates.
5. Grow cells until density reaches 90 % confluency per well.
6. To activate the Rac/p38 MAPK pathway, remove growth media and add 100 μL of hyperosmotic media containing 175 mM NaCl. Add fresh growth media to control wells to visualize ground state (*see* Fig. 2).
7. To visualize inhibition of the translocation, add compounds to a final concentration of 15 μM to experimental wells containing hyperosmotic media.
8. Add DMSO to no-inhibitor control wells. Mix the plates carefully and incubate them at 37 °C in a humidified chamber atmosphere containing 5 % CO_2.
9. Stop the treatment after 70–90 min by disposing of the medium.
10. After removal of the medium wash the cells two times with 100 μL PBS.
11. For fixation of the cells, add 50 μL of the 4 % fixation solution. Incubate for 15 min at room temperature.
12. Remove the fixation solution and wash the cells with PBS.
13. For staining of the nuclei, solubilize cellular membranes by washing the cells two times with PBST. Add 100 μL DAPI solution and incubate for 10 min at room temperature in the dark.
14. Remove DAPI solution by washing the cells three times with PBS.
15. Cover the cells with 100 μL PBS and store the plates at 4 °C until ready for imaging.

3.2.2 Image Acquisition

1. Acquire images of EGFP and DAPI fluorescent cells using a high-content analysis imaging platform (e.g., ImageXpress micro XL microscope, Molecular Devices).
2. For image acquisition use a 20× Super Pan fluor ELWD cc 0–2 mm objective in combination with an appropriate filter set (e.g., EGFP: 482 ± 35 nm; DAPI: 377 ± 50 nm).

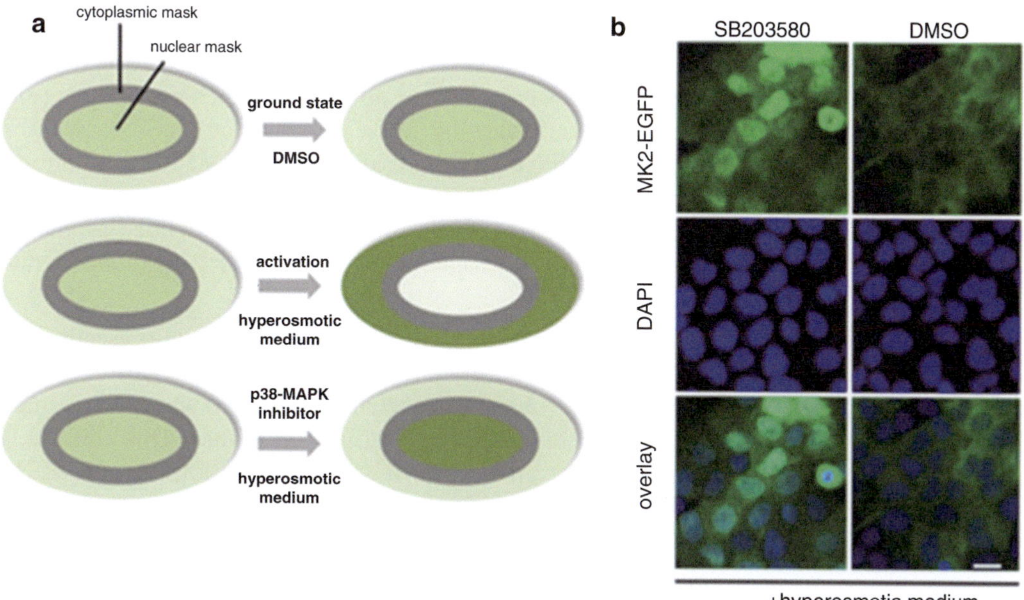

Fig. 2 Cellular MK2-EGFP translocation assay to screen for p38 inhibitors. (**a**) Schematical outline of MK2-EGFP translocation assay. Nuclear DAPI segmentation to determine the nucleocytoplasmic MK-2-EGFP fluorescence intensity ratio (*grey ring*). In the ground state (U2OS cells stably expressing MK2-EGFP), a slightly higher fluorescent intensity in the nucleus compared to the cytoplasm will be detected. For activation of the Rac/p38 pathway, cells were subjected to hyperosmotic medium (175 mM NaCl), leading to a rapid translocation of MK2-EGFP from the nucleus to the cytoplasm. Co-treatment with p38 inhibitors abolishes this redistribution. (**b**) Representative pictures of activated U2OS MK2-EGFP cells, either non-treated (DMSO) or treated with p38 inhibitor SB203580; scale bar 20 μM

3. Set appropriate exposure times (e.g., 4 ms for DAPI and 110 ms for EGFP channel). To allow reliable statistical analysis and quantification, at least 400–500 cells per well (corresponding to individual compound treatment) should be imaged.

3.2.3 Data Assessment

1. Determination of ratios of nucleocytoplasmic fluorescent intensities with automated image acquisition: For analysis of the translocation of the fluorescent signal of MK2-EGFP upon activation of p38-MAPK, a nuclear compartment surrounded by a peripheral DAPI-negative cytoplasmic ring has to be defined based on a translocation-enhanced application module.
2. Use the following settings: DAPI nuclear mask with Vernier-adjustments: compartment image: DAPI; translocation probe image: EGFP; compartment algorithm: standard; approximate width: 13 μm; intensity above local background: 656–65535 gray levels; minimum area: 60 pixel; maximum area: 355 pixel; auto-separate touching objects; inner region distance: 0.5 μm;

outer region distance: 0.5 μm; outer region width: 2 μm; background estimation method: auto-constant.

3. For image analysis, use an automated software analysis (e.g., Met-aXpress, Molecular Devices). Calculate fluorescent intensity ratios using the average of the cellular MK2-EGFP fluorescence intensity (Cell: nu-clear inner/outer mean intensity) of all images/sites per well. p38 activation is defined as positive when the nucleocytoplasmic MK2-fluorescence intensity ratio is below 0.7.
4. Definition of positive and negative controls: Fluorescent ratios of cells treated with DMSO are defined as negative controls (MK2-EGFP translocates to the cytoplasm) whereas treatment with an established p38-MAPK inhibitor (SB203580) is defined as positive control (upon inhibition of p38-MAPK MK2-EGFP retains in the nucleus); *see* Fig. 2.
5. Determination of Z' factors

   Z' factors were calculated according to the formula

   $z' = 1 - (3(\theta_p + \theta_n)/(\mu_p - \mu_p))$

   p: positive control and n: negative control, θ: standard deviation, μ: mean [13].

3.3 Stimulation of TNFα Release from Fresh Human Whole Blood of Two Different Donors While Incubating p38 MAPK Inhibitors

All compounds will be incubated in the blood of two different blood donors. Arrangement for the dilution procedure of seven test compounds and one reference (e.g., SB 203580) is described here.

In order to obtain the working dilutions out of the 10 mM stock solutions of seven test compounds and reference, a dilution row is prepared by tenfold serial dilution. The lowest concentration of the dilution row should contain a tenfold excess of the final assay concentration to be incubated in the whole blood. For test compounds with potencies similar to the reference compound SB 203580, the lowest concentration of the dilution row is 0.1 μM. The subsequent addition of these working dilutions to the human whole blood will generate another tenfold diminution of the concentration and the final assay concentration is achieved.

The concentration of Cr-EL/EtOH cannot exceed 1 % in those dilutions which will be subsequently added to the human whole blood.

3.3.1 Compound Dilution and Incubation of Whole Blood

1. Separate 1.5 mL Eppendorf reaction tubes for preparation of the compound/reference dilutions in an appropriate support. For seven test compounds and the reference, 40 reaction tubes will be needed.
2. Dispense 450 μL of DPBS-gentamicin buffer in the reaction tubes for the dilutions of 1 mM and 100 μM.

3. Dispense 450 μL of 1 % Cr-EL/EtOH in DPBS-gentamicin buffer in the reaction tubes for the dilutions of 10, 1, and 0.1 μM.
4. Dilution procedure of test compounds and reference is carried out for each compound according to the scheme described in Table 2.
5. Dilutions containing 1 % Cr-EL/EtOH are appropriate for incubation in the human whole blood and will be transferred to a 96-well Masterplate (see next step). The dilutions containing higher amounts of Cr-EL/EtOH are discarded.
6. Transfer 300 μL of those test compound/reference dilutions containing 1 % Cr-EL/EtOH into a 96-well Masterplate according to the scheme in Table 3 (*see* **Note 5**).
7. Dispense 300 μL of 1 % Cr-EL/EtOH in DPBS-gentamicin buffer into wells for basal and STIM controls. Basal control corresponds to basal TNFα concentration in unstimulated whole blood. STIM controls correspond to maximal TNFα concentration in whole blood after stimulation with 0.1 mg/mL LPS-Solution.
8. The FBS is carefully thawed. One hour prior to its use, the FBS is brought to an appropriate working temperature in the incubator at 37 °C.
9. Pipet 16 mL of FBS into two 50 mL Falcon tubes labeled for blood donors 1 and 2.
10. From both donors, more than 16 mL of fresh potassium EDTA-anticoagulated human whole blood in a blood collection tube is needed. The whole blood will be diluted 1:2 into the FBS inside the labeled 50 mL Falcon tube as follows.

Table 2
Overview of the testing compound/reference dilution procedure

Concentration of test compound (μM)	Concentration of Cr-EL/EtOH (%)	Dilution
1000	10	450 μL DPBS-gentamicin buffer + 50 μL 10 mM stock solution
100	1	450 μL DPBS-gentamicin buffer + 50 μL 1000 μM stock solution
10	1	450 μL 1 % Cr-EL/EtOH in DPBS-gentamicin buffer + 50 μL 100 μM stock solution
1	1	450 μL 1 % Cr-EL/EtOH in DPBS-gentamicin buffer + 50 μL 10 μM stock solution
0.1	1	450 μL 1 % Cr-EL/EtOH in DPBS-gentamicin buffer + 50 μL 1 μM stock solution

11. With the help of a pipet with appropriate tip, four times 1 mL of blood is slowly added to the FBS.
12. After each addition of 4 mL of blood, the blood collection tube is closed and pivoted carefully. The pipet tip is exchanged and this process is repeated until 16 mL of whole blood is added to the FBS.
13. After the complete addition of the blood to the Falcon tube, the contents are mixed by careful pivoting the tube and then it is stored in the incubator at 37 °C until use (*see* **Note 12**).
14. Repeat this procedure with the whole blood of donor 2.
15. Transfer 400 μL of the FBS-diluted whole blood of donors 1 and 2 to the deepwell plate. One deepwell plate carries the diluted blood of two donors, according to the arrangement shown in Fig. 3.
16. From the compound dilutions Masterplate prepared in **steps 6** and **7** (*see* Table 3), transfer 50 μL of the test compounds/reference dilutions into the diluted blood samples of both donors following the same scheme. For the wells of basal and STIM controls, 50 μL 1 % Cr-EL/EtOH in DPBS-gentamicin buffer is transferred (*see* **Note 13**).
17. Preincubate the deepwell plate containing the diluted human whole blood with the test compounds and reference dilutions for 15 min at 37 °C in the CO_2 incubator.

3.3.2 Stimulation of TNFα Release from the Diluted Human Whole Blood

1. After 15 min of preincubation, the deepwell plate containing the diluted whole blood and sample dilutions is taken out of the incubator.

Table 3
Pipetting scheme of the Masterplate with test compound/reference dilutions

	B 10 μM	D 100 μM	E 0.1 μM	G 1 μM
Basal	B 1 μM	D 10 μM	F 100 μM	G 0.1 μM
STIM 1	B 0.1 μM	D 1 μM	F 10 μM	H 100 μM
A 100 μM	**STIM 2**	D 0.1 μM	F 1 μM	H 10 μM
A 10 μM	C 100 μM	**STIM 3**	F 0.1 μM	H 1 μM
A 1 μM	C 10 μM	E 100 μM	**STIM 4**	H 0.1 μM
A 0.1 μM	C 1 μM	E 10 μM	G 100 μM	**STIM 5**
B 100 μM	C 0.1 μM	E 1 μM	G 10 μM	

Basal = Basal TNFα concentration in unstimulated whole blood
STIM = Maximal TNFα concentration in whole blood after stimulation with 0.1 mg/mL LPS
A–H = Dilution wells of seven test compounds and one reference, four concentrations each

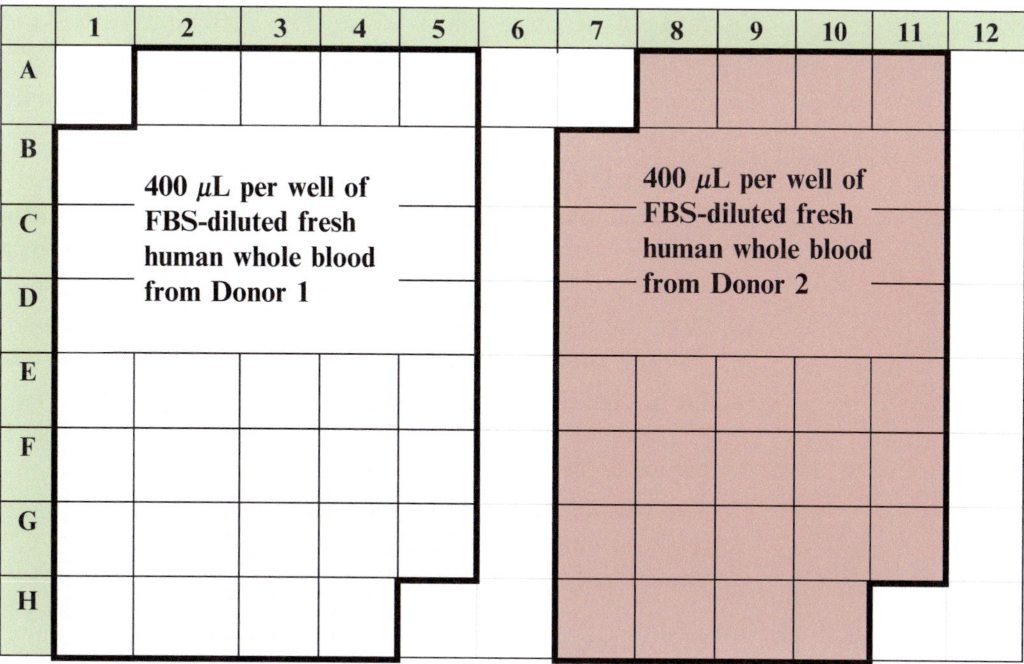

Fig. 3 Arrangement of the deepwell plate

2. Preset the LPS solution in a suitable container, which allows simultaneous sampling with all channels of the eight-channel pipet (*see* **Note 14**).
3. Add 50 μL of LPS solution to all wells of the deepwell plate that contain blood, except the wells for basal.
4. For basal control wells, 50 μL of 0.025 % DPBS-gentamicin buffer are added separately.
5. Incubate the deepwell plate at 37 °C incubator with 6 % CO_2 for 2.5 h.
6. To prepare for termination of sample incubation and TNFα release in the diluted blood samples, separate and label 76 Eppendorf reaction tubes for each donor, as duplicate aliquots from each blood-containing well will be sampled.
7. Place the Eppendorf reaction tubes in an appropriate support that allows ice cooling and for the deepwell plate, prepare an ice-water mixture in suitable container.
8. Cool the 1 % BSA/PBS buffer on ice.
9. Switch on the high-speed centrifuge to ensure the attainment of a working temperature of 4 °C.
10. Termination of TNFα release: After the incubation period, remove the deepwell plate from the incubator and place into the container of ice-water mixture. The plate is kept in this container until centrifugation.

11. Add 500 μL of ice-cold 1 % BSA/PBS buffer into each blood-containing well.
12. The deepwell plate is left to rest for 10 min inside the ice-water mixture.
13. Seal the deepwell plate carefully with a cap mat.
14. For separation of TNFα-containing plasma supernatant from cellular components, centrifuge the sealed deepwell plate at 1000 × *g* and 4 °C for 15 min in the high-speed centrifuge.
15. Take the deepwell plate from the high-speed centrifuge and carefully remove the seal.
16. From each well, two plasma aliquots of 160 μL each are sampled into the previously provided Eppendorf reaction tubes in the ice-cooled support (*see* **Note 15**).
17. The plasma aliquots are immediately frozen at −20 °C until ready to quantify TNFα concentration. Storage at −20 °C is limited to 1 month.

3.3.3 ELISA for TNFα Quantification

1. The plasma samples from both blood donors 1 and 2 will be analyzed in duplicate. Thus, two Nunc™ Maxisorp plates are needed, corresponding to blood donors 1 and 2.
2. Prepare an appropriate amount of working solution of the capture antibody (human TNFα DuoSet) according to the manufacturer's instructions.
3. To coat the plates, pipet 100 μL of capture antibody working solution into each well of both Nunc™ Maxisorp plates.
4. Cover the plates with a lid for adsorption of the capture antibody to the plates' surface overnight at room temperature in the dark.
5. The next day, discard the coating solution from the first assay plate. Wash the plate three times with the help of a wash buffer-containing wash bottle (*see* **Note 16**).
6. Pat the plate onto an adsorptive surface to remove residual wash buffer.
7. Repeat this procedure for the second plate corresponding to blood donor 2.
8. Blocking step: 300 μL of 1 % BSA/PBS buffer is pipetted into each well of the assay plates.
9. The plates are covered with a lid and incubated for 1 h at room temperature. During this 1-h blocking step, the calibration row of the recombinant human TNFα standard and the dilution of the plasma samples are performed.
10. For both assay plates, separate eight 1.5 mL Eppendorf reaction tubes in a support that allows ice cooling. The reaction

tubes are labeled according to Table 4. The following procedure is repeated to result in two calibration rows, one for each of the blood donor assay plate.

11. To prepare the calibration row of the recombinant human TNFα standard, add the corresponding amounts of 1 % BSA/PBS buffer from Table 4 into the Eppendorf reaction tubes (S2–S8).
12. The reconstituted TNFα standard is thawed at room temperature and then left on ice to avoid potential loss of TNFα. Depending on the concentration of the reconstituted TNFα standard, the amount of TNFα standard which is needed to result in a concentration of 1000 pg/mL of TNFα has to be calculated (Tube S1).
13. Before starting the dilution series the reconstituted TNFα standard solution is vortexed thoroughly for 25–30 s.
14. From the tube containing the concentrated reconstituted TNFα standard (1000 pg/mL) transfer 300 μL into the first Eppendorf reaction tube containing the respective volume of 1 % BSA/PBS buffer (*see* Table 4). The tip is rinsed three to five times in the solution. Afterwards, the dilution is vortexed for approximately 30 s.
15. Further dilution steps are performed by pipetting 300 μL of the previous dilution into the next Eppendorf reaction tube containing 300 μL of 1 % BSA/PBS (*see* Table 4). The tip is rinsed three to five times in the final solution. Afterwards it is vortexed for approximately 30 s.
16. Do not dilute into the tube marked S8. This tube will contain only 300 μL of 1 % BSA/PBS buffer.

Table 4
Dilution procedure for the TNFα calibration row

Tube #	TNFα concentration (pg/mL)	Reconstituted TNFα standard dilution (μL)	1 % BSA/PBS buffer (μL)
S1	1000	See manufacturer's indications	q.s.
S2	500	300 from **S1**	300
S3	250	300 from **S2**	300
S4	125	300 from **S3**	300
S5	62.5	300 from **S4**	300
S6	31.25	300 from **S5**	300
S7	15.6	300 from **S6**	300
S8	0	–	300

17. The dilutions of the TNFα standard calibration row remain on ice until use.
18. To dilute the plasma samples with TNFα-diluent 2, from each blood donor, only one aliquot of TNFα-containing plasma samples is thawed for immediate analysis. The other aliquot remains in the freezer as a backup.
19. The thawed plasma samples remain in an ice-cooled support to avoid a temperature-derived loss of TNFα. Repeated freeze-thaw cycles have to be avoided under any circumstances.
20. Reconstitute the TNFα-diluent 2. Approximately 7 mL of TNFα-diluent 2 will be needed for one experiment.
21. As soon as the plasma samples are thawed, they have to be immediately stored on ice for the further dilution procedure.
22. Starting with the plasma samples of blood donor 1, add 80 μL of TNFα-diluent 2 into each Eppendorf reaction tube.
23. Repeat this procedure for the plasma samples of donor 2.
24. All diluted plasma samples are vortexed for 25 s.
25. Until being transferred to the respective assay plate, the diluted plasma samples are stored again on ice. This transfer should follow promptly.
26. When blocking step (**step 8**) is over, discard blocking solution from the assay plate corresponding to donor 1. Wash the plate three times with the help of a wash buffer-containing wash bottle.
27. Pat the plate onto an adsorptive surface to remove residual wash buffer.
28. Repeat this procedure for the second plate corresponding to blood donor 2 (*see* **Note 16**).
29. Starting with the highest concentration of 1000 pg/mL, 50 μL of each solution of the TNFα standard calibration row is transferred in duplicate to the wells of the assay plate of donor 1 according to the scheme in Fig. 4.
30. The wells A3 and A4 remain empty, as they will serve to determine blank values.
31. 50 μL of the diluted plasma samples of blood donor 1 are pipetted in duplicates from the Eppendorf reaction tubes to the wells of the corresponding assay plate according to Fig. 4.
32. Repeat this procedure for TNFα standard calibration row and plasma samples of donor 2.
33. Cover the assay plates with a lid and incubate them for 2 h in darkness at room temperature.
34. Fifteen minutes prior to the end of the incubation time of TNFα standard calibration row and diluted plasma samples,

	1	2	3	4	5	6	7	8	9	10	11	12
A	S 1	S 1	Blank	Blank	B 1μM	B 1μM	D 10μM	D 10μM	E 0.01μM	E 0.01μM	G 0.1μM	G 0.1μM
B	S 2	S 2	Basal	Basal	B 0.1μM	B 0.1μM	D 1μM	D 1μM	F 10μM	F 10μM	G 0.01μM	G 0.01μM
C	S 3	S 3	Stim 1	Stim 1	B 0.01μM	B 0.01μM	D 0.1μM	D 0.1μM	F 1μM	F 1μM	H 10μM	H 10μM
D	S 4	S 4	A 10μM	A 10μM	Stim 2	Stim 2	D 0.01μM	D 0.01μM	F 0.1μM	F 0.1μM	H 1μM	H 1μM
E	S 5	S 5	A 1μM	A 1μM	C 10μM	C 10μM	Stim 3	Stim 3	F 0.01μM	F 0.01μM	H 0.1μM	H 0.1μM
F	S 6	S 6	A 0.1μM	A 0.1μM	C 1μM	C 1μM	E 10μM	E 10μM	Stim 4	Stim 4	H 0.01μM	H 0.01μM
G	S 7	S 7	A 0.01μM	A 0.01μM	C 0.1μM	C 0.1μM	E 1μM	E 1μM	G 10μM	G 10μM	Stim 5	Stim 5
H	S 8	S 8	B 10μM	B 10μM	C 0.01μM	C 0.01μM	E 0.1μM	E 0.1μM	G 1μM	G 1μM		

Fig. 4 Arrangement of TNFα standard calibration row and plasma samples on the assay plate. Basal corresponds to basal TNFα concentration in unstimulated whole blood. STIM, Maximal TNFα concentration in whole blood after stimulation with 0.1 mg/mL LPS. A–H = seven test compounds and one reference-treated plasma samples, four concentrations each

prepare the working solution of the biotinylated caprine anti-human TNFα detection antibody, according to the manufacturer's instructions.

35. Discard the TNFα standard and diluted plasma samples from the assay plate corresponding to donor 1. Wash the plate three times with the help of a wash buffer-containing wash bottle.
36. Pat the plate onto an adsorptive surface to remove residual wash buffer.
37. Repeat this procedure for the second plate corresponding to blood donor 2 (*see* **Note 16**).
38. Pipet 100 μL of the working solution of the TNFα detection antibody into each well of the assay plate corresponding to donor 1.
39. The blank wells A3 and A4 remain empty.
40. Repeat this procedure for the second plate corresponding to blood donor 2.
41. Cover the assay plates with a lid and incubate with TNFα detection antibody for 2 h in darkness at room temperature.

42. Twenty minutes prior to the end of the incubation time of TNFα detection antibody, take the TMB substrate reagent pack out of the fridge. For tempering, insert 11 mL of both solutions A and B into separate 15 mL Falcon tubes. Store both solutions in a dark place.
43. Fifteen minutes prior to the end of the incubation time of TNFα detection antibody, prepare the working solution of the enzyme reagent, according to the manufacturer's instructions.
44. When the incubation of TNFα detection antibody is over, discard the TNFα detection antibody from the assay plate corresponding to donor 1.
45. Wash the plate three times with the help of a wash buffer-containing wash bottle.
46. Pat the plate onto an adsorptive surface to remove residual wash buffer.
47. Repeat this procedure for the second plate corresponding to blood donor 2 (*see* **Note 16**).
48. Pipet 100 μL of the working solution of the enzyme reagent into each well of the assay plate corresponding to donor 1.
49. The blank wells A3 and A4 remain again empty.
50. Repeat this procedure for the second plate corresponding to blood donor 2.
51. Cover the assay plates with a lid and incubate with the enzyme reagent for 20 min in darkness at room temperature.
52. Twenty minutes prior to the end of the incubation time of enzyme reagent, mix the now tempered TMB substrate reagent solution A and B into one 50 mL Falcon tube. Keep the solution in the dark.
53. When the incubation of the enzyme reagent is over, discard the enzyme reagent from the assay plate corresponding to donor 1.
54. Wash the plate three times with the help of a wash buffer-containing wash bottle.
55. Pat the plate onto an adsorptive surface to remove residual wash buffer.
56. Repeat this procedure for the second plate corresponding to blood donor 2 (*see* **Note 16**).
57. 100 μL of the pooled TMB substrate reagent solutions are pipetted into all wells of the assay plate corresponding to blood donor 1.
58. Repeat this procedure for the second plate corresponding to blood donor 2 (*see* **Note 17**).

59. Incubate the uncovered assay plates with the TMB substrate reagent for 20 min in darkness.
60. During the incubation of the TMB substrate reagent, a deep sky blue coloring develops.
61. To terminate the color reaction, add 50 μL of sulfuric acid into all wells of the assay plate corresponding to blood donor 1.
62. Repeat this procedure for the second plate corresponding to blood donor 2 (*see* **Note 17**).
63. Both assay plates are read out as fast as possible using the EMax Precision Microplate ELISA reader.

3.3.4 ELISA Data Assessment

TNFα concentration in the plasma derived from whole blood of two different donors as an indicator of p38 MAPK inhibitor efficacy is determined after readout of the optical densities at 450 nm.

Quantification of TNFα concentration is performed using the TNFα standard calibration row.

The mean optical densities of the wells for basal TNFα level serve as negative control and are used to reassess the quality of the experimental data. They provide information about elevated TNFα levels, which can be donor specific or provoked by mishandling of the blood samples during pipetting or transfer procedures. Ideally, basal TNFα level will be below the range of the TNFα standard calibration row and can be regarded as zero.

The mean optical densities of the STIM wells serve as positive control and are used to reassess the quality of LPS-induced TNFα release from the blood samples. They provide information about the inducibility of TNFα release, which can be prevented by the intake of anti-inflammatory drugs (e.g., NSAIDs) or by a non-response of the blood samples upon LPS stimulation.

Calculation is carried out separately for each donor.

1. The mean optical densities of the blank wells (A3 and A4) are subtracted from the optical density of each well.
2. Calculate the relation between TNFα concentration [pg/mL] and optical density with the help of TNFα standard calibration row.
3. Calculate the mean TNFα concentration [pg/mL] of the basal wells. Ideally, basal TNFα level will be below the range of the TNFα standard calibration row and can be regarded as zero.
4. Calculate the mean TNFα concentration [pg/mL] of the STIM value. Ideally, maximum TNFα levels of >300 pg/mL are achieved.
5. The mean TNFα level of the STIM values corresponds to maximum TNFα release and is accordingly regarded as 0 % inhibition.

6. Percentage of inhibition is calculated from the mean TNFα concentration of the test compound or reference with respect to the TNFα levels achieved for STIM according to the following equation:

$$\text{Inhibition}[\%] = 100 - \left(\frac{\text{TNF– conc. Sample}}{\text{TNF– conc. STIM}}\right) \times 100$$

4 Notes

1. The use of a reference compound is advisable, as the obtained IC_{50} value of the reference compound allows the evaluation of the accuracy of the experiment. The IC_{50} value of the reference SB203580 should be approximately 0.05 μM ± 0.001 [6].
2. Since also other solutions are produced with PBS buffer, preparation of a bigger volume should be considered.
3. This solution will also be used in the ELISA part. Strictly avoid contamination!
4. Since LPS is not readily soluble and a complete dissolution has to be ensured an ultrasonic bath can be used to accelerate dissolution.
5. If there has been a long time span between diluting and transferring, the testing compound dilutions should be vortexed thoroughly.
6. Coating can also be carried out at the day of the experiment. Instead of the overnight procedure, the coating solution is incubated for 90 min at 37 °C. Anyway, better consistency between the triplicates will be achieved when coating is performed overnight.
7. The wells are filled with wash buffer with a strong jet, after which the wash buffer from the plate is discarded vigorously. This process is repeated three times.
8. Ensure that the inserted solution is in contact with the surface of the well. Remove potential air bubbles by carefully tapping against one side of the plate.
9. Depending on the inhibitory potency, this concentration can be disregarded. For test compounds with potencies in the range of the reference SB 203580, this solution will not be transferred to the plate.
10. The mean optical density of the NSB wells should not exceed a value of 0.1.
11. The mean optical density of the STIM wells should reach the range of 0.5–0.7.

12. The blood is pipetted very carefully, and down the rim of the Falcon tube.
13. The addition of the test compounds/reference dilutions to the diluted whole blood results in a further tenfold dilution! Therefore, the test compounds and reference should only be diluted to a tenfold excess of the final assay concentration prior to the addition into the diluted blood!
14. The tips of the pipet should be skimmed off at the rim of the trough before addition of the LPS solution into the compound-containing whole blood of the deepwell plate.
15. The plasma supernatant is carefully sampled by pipetting alongside the side of the well. To avoid swirling up the cellular components from the bottom, the pipet tip is placed at half height of the well. The sampled plasma should be a clear solution.
16. The wells are filled with wash buffer with a strong jet, after which the wash buffer from the plate is discarded vigorously. This process is repeated three times.
17. The solution has to be transferred as quickly as possible to decrease the shift due to a bigger difference in reaction time.

Acknowledgement

The authors would like to thank Stefanie Mocka, Carolin Klein, Anke Friedrichs, Dr. Sabine Luik, Dr. Sabine Linsenmeier, Dr. Cornelia Greim, and K. Bauer for their valuable contributions to the establishment and optimization of the herein presented whole blood assay.

UR gratefully acknowledges the Ministry of Science, Research and Arts of Baden-Württemberg (V.1.4.-H3-1403-74) for financial support.

References

1. Coulthard LR et al (2009) p38(MAPK): stress responses from molecular mechanisms to therapeutics. Trends Mol Med 15(8):369–379
2. Korb A et al (2006) Differential tissue expression and activation of p38 MAPK alpha, beta, gamma, and delta isoforms in rheumatoid arthritis. Arthritis Rheum 54(9): 2745–2756
3. Schett G, Zwerina J, Firestein G (2008) The p38 mitogen-activated protein kinase (MAPK) pathway in rheumatoid arthritis. Ann Rheum Dis 67(7):909–916
4. Zarubin T, Han JH (2005) Activation and signaling of the p38 MAP kinase pathway. Cell Res 15(1):11–18
5. Kumar S, Boehm J, Lee JC (2003) p38 map kinases: key signalling molecules as therapeutic targets for inflammatory diseases. Nat Rev Drug Discov 2(9):717–726
6. Goettert M, Graeser R, Laufer SA (2010) Optimization of a nonradioactive immunosorbent assay for p38alpha mitogen-activated protein kinase activity. Anal Biochem 406(2):233–234
7. Anton R et al (2014) A p38 substrate-specific MK2-EGFP Translocation assay for identification and validation of new p38 inhibitors in living cells: a comprising alternative for acquisition of cellular p38 inhibition data. PLoS One 9(4):e95641

8. Engel K, Kotlyarov A, Gaestel M (1998) Leptomycin B-sensitive nuclear export of MAPKAP kinase 2 is regulated by phosphorylation. EMBO J 17(12):3363–3371
9. Williams RG et al (2006) Generation and characterization of a stable MK2-EGFP cell line and subsequent development of a high-content imaging assay on the Cellomics ArrayScan platform to screen for p38 mitogen-activated protein kinase inhibitors. Methods Enzymol 414:364–389
10. Trask OJ et al (2009) High-throughput automated confocal microscopy imaging screen of a kinase-focused library to identify p38 mitogen-activated protein kinase inhibitors using the GE InCell 3000 analyzer. Methods Mol Biol 565:159–186
11. Ross S et al (2006) High-content screening analysis of the p38 pathway: profiling of structurally related p38alpha kinase inhibitors using cell-based assays. Assay Drug Dev Technol 4(4):397–409
12. Thurm CW, Halsey JF (2005) Measurement of cytokine production using whole blood. Curr Protoc Immunol Chapter 7: p. Unit 7 18B
13. Zhang JH, Chung TDY, Oldenburg KR (1999) A simple statistical parameter for use in evaluation and validation of high throughput screening assays. J Biomol Screen 4(2): 67–73

Chapter 11

Genetically Encoded Fluorescent Indicators to Visualize Protein Phosphorylation in Living Cells

Moritoshi Sato and Yoshio Umezawa

Abstract

Protein phosphorylation by intracellular kinases plays one of the most pivotal roles in signaling pathways within cells. To reveal the biological processes related to the kinase proteins, electrophoresis, immunocytochemistry, and in vitro kinase assay have been used. However, these conventional methods do not provide enough information about spatial and temporal dynamics of the signal transduction based on protein phosphorylation and dephosphorylation in living cells. To overcome the limitation for investigating the kinase signaling, we developed genetically encoded fluorescent indicators for visualizing the protein phosphorylation in living cells. Using these indicators, we visualized under a fluorescence microscope when, where, and how the protein kinases are activated in single living cells.

Key words Fluorescent indicators, Fluorescence resonance energy transfer (FRET), Live cell imaging, Protein phosphorylation reactions, Protein kinases, Subcellular dynamics

1 Introduction

To visualize signal transduction based on protein phosphorylation reactions in living cells, we have developed genetically encoded fluorescent indicators (Fig. 1) [1–6]. A substrate domain for a protein kinase of interest is fused with a phosphorylation recognition domain via a flexible linker sequence. The tandem fusion unit consisting of the substrate domain, linker sequence, and phosphorylation recognition domain is sandwiched between two different colored fluorescent proteins, enhanced cyan fluorescent protein (ECFP), and enhanced yellow fluorescent protein (EYFP), which serve as the donor and acceptor fluorophores for fluorescence resonance energy transfer (FRET), respectively. As a result of phosphorylation of the substrate domain and the subsequent binding of the phosphorylated substrate domain with the adjacent phosphorylation recognition domain, FRET is induced between the two fluorescent units, which elicits phosphorylation-dependent changes in fluorescence emission ratios of the donor and acceptor fluorescent

Hicham Zegzouti and Said A. Goueli (eds.), *Kinase Screening and Profiling: Methods and Protocols*, Methods in Molecular Biology, vol. 1360, DOI 10.1007/978-1-4939-3073-9_11,

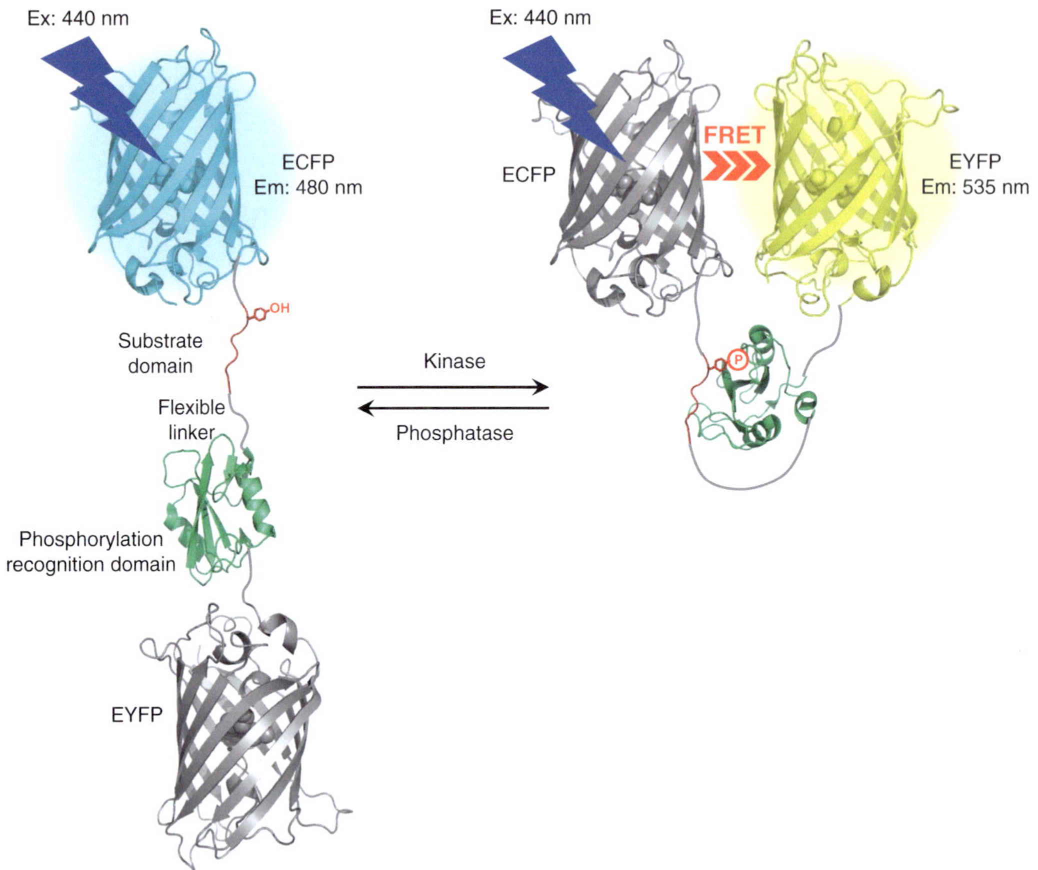

Fig. 1 Principle of the present fluorescent indicator for protein phosphorylation, which was named phocus. Upon phosphorylation of the substrate domain within phocus by the protein kinase, the adjacent phosphorylation recognition domain binds with the phosphorylated substrate domain, which changes the efficiency of FRET between the GFP mutants (ECFP and EYFP) within)phocus. By tethering a subcellular localization domain, phocus can be localized in a subcellular compartment of interest to visualize phosphorylation reactions there. *ECFP* enhanced cyan fluorescent protein, *EYFP* enhanced yellow fluorescent protein, *OH* hydroxy group, P in an *open circle*, phosphorylated hydroxy group

proteins. Upon activation of protein phosphatases, the phosphorylated substrate domain is dephosphorylated, decreasing the FRET signal of the fluorescent indicator. We named this type of fluorescent indicator "phocus" (fluorescent indicator for protein *pho*sphorylation that can be *cus*tom-made). By using suitable substrate and phosphorylation recognition domains, we have developed fluorescent indicators for several key protein kinases, such as a receptor tyrosine kinase (insulin receptor) [1], a non-receptor tyrosine kinase (c-Src) [2, 3], and serine/threonine protein kinases (Akt/PKB [4], ERK [5], and JNK [6]) (Table 1). In addition, these fluorescent indicators have been further tailored to visualize subcellular localization of protein kinase activity in living cells by fusing appropriate subcellular localization sequences/domains.

Table 1
Detailed information about fluorescent indicators to visualize protein phosphorylation reactions in living cells

Kinase	Indicator	Subcellular localization domain/sequence	Localization	Substrate domain	Phosphorylation recognition domain
Insulin receptor (Tyr kinase)	Phocus-2pp	PH-PTB domain from IRS-1	Cytosol, nucleus and membrane ruffles	ETGTEEYMKMDJG	SH2 domain ($p85a_{330-429}$)
c-Src (Tyr kinase)	Srcus	–	Cytosol	EEEIYGEFF	SH2 domain ($c\text{-}Src_{148-248}$)
	PTB-Srcus	Shc_{55-206}	Cytosol and EGFR	EEEIYGEFF	SH2 domain ($c\text{-}Src_{148-248}$)
	TM-Srcus	Cbp_{1-52}	Plasma membrane	EEEIYGEFF	SH2 domain ($c\text{-}Src_{148-248}$)
Akt/PKB (Ser/Thr kinase)	Aktus	–	Cytosol	RGRSRSAP	$14\text{-}3\text{-}3\eta_{82-235}$
	eNOS-Aktus	$eNOS_{1-35}$	Golgi	RGRSRSAP	$14\text{-}3\text{-}3\eta_{82-235}$
	Bad-Aktus	$Tom20_{1-33}$	Mitochondrial outer membrane	RGRSRSAP	$14\text{-}3\text{-}3\eta_{82-235}$
ERK (Ser/Thr kinase)	Erkus	–	Cytosol and nucleus	KRELVEPLTPSIEAPNQALLR	FHA2 domain from Rad53p
	Erkus-nuc	NLS	Nucleus	KRELVEPLTPSIEAPNQALLR	FHA2 domain from Rad53p
	Erkus-cyto	NES	Cytosol	KRELVEPLTPSIEAPNQALLR	FHA2 domain from Rad53p
JNK (Ser/Thr kinase)	JuCKY	–	Cytosol	DPVGSLKPHLRAKNSDLLTAPDVGLLKLATPELERL	FHA2 domain from Rad53p
	JuCKY-NLS	NLS	Nucleus	DPVGSLKPHLRAKNSDLLTAPDVGLLKLATPELERL	FHA2 domain from Rad53p

2 Materials

2.1 Cell Culture and Transfection

1. Chinese hamster ovary cells expressing insulin receptor (CHO-IR cells).
2. Ham's F-12 medium (Life Technologies).
3. Fetal bovine serum (Life Technologies).
4. Trypsin EDTA (Life Technologies).
5. LipofectAMINE 2000 (Life Technologies).
6. Glass-bottomed dish (Asahi Techno Glass, Japan)
7. pcDNA 3.1(+) vector (Life Technologies).

2.2 Live Cell Imaging

1. Hank's balanced salt solution (Life Technologies).
2. Excitation filter for ECFP (440AF21) (Omega Optical Inc.).
3. Dichroic mirror (455DRLP) (Omega Optical Inc.).
4. Emission filter for ECFP (480AF30) (Omega Optical Inc.).
5. Emission filter for EYFP (535AF26) (Omega Optical Inc.).
6. Insulin (Sigma-Aldrich).

3 Methods

3.1 Constructing cDNA Encoding the Fluorescent Indicator

1. To construct the fluorescent indicator, named phocus-2pp, fragment cDNAs encoding ECFP, EYFP, substrate domain, phosphorylation recognition domain, and pleckstrin homology-phosphotyrosine-binding (PH-PTB) domain were generated by standard polymerase chain reaction (PCR) to create restriction sites.
2. ECFP and EYFP are different colored variants of green fluorescent protein (GFP) derived from *Aequorea victoria* and optimized for mammalian codons with the following additional mutations: ECFP mutations are F64L/S65T/Y66W/N146I/M153T/V163A/N212K; EYFP mutations are S65G/V68L/Q69K/S72A/T203Y.
3. The amino acid sequence of the substrate domain is ETGTEEYMKMDLG, which is derived from the insulin receptor substrate-1 (IRS-1) and phosphorylated by insulin receptor.
4. The phosphorylation recognition domain is an N-terminal SH2 domain (SH2n) derived from p85α subunit ($p85\alpha_{330-429}$) of phosphatidylinositol 3-kinase, which binds to the phosphorylated substrate domain derived from IRS-1.
5. We used the amino acid sequence of GNNGGNNNGGS for a flexible linker that links the substrate domain with the phosphorylation recognition domain.

6. The PH and PTB domain are derived from an N-terminal domain of IRS-1 (IRS-1_{1-271}) and bind to the phosphoinositides at the plasma membrane and the juxtamembrane domain of insulin receptor, which is immediately tyrosine-phosphorylated by insulin stimulation, respectively. Thus, tagging with the PH-PTB domain increases the concentration of phocus-2pp around the insulin receptor at the plasma membrane upon insulin stimulation. This would enhance the phosphorylation kinetics of the fluorescent indicator, thereby facilitating the measurement of protein phosphorylation reactions.
7. Although we do not show it in this chapter, adding a nuclear-export-signal sequence (NES) is useful in ensuring that the fluorescent indicator is excluded from the nucleus. For this purpose, we often use the amino acid sequence LPPLERLTL, which is derived from a human immunodeficiency virus-derived protein, Rev.
8. The full cDNA encoding phocus-2pp was subcloned at *Hind* III and *Xba* I sites of a mammalian expression vector, pcDNA3.1 (+).
9. The cDNA encoding phocus-2pp was biologically amplified by transforming *E. coli* with the plasmid, and purified with a commercial plasmid purification kit.

3.2 Cell Culture and Transfection

1. Plate CHO-IR cells on glass-bottomed dishes with Ham's F-12 medium supplemented with 10 % fetal bovine serum.
2. Incubate the cells at 37 °C in 5 % CO_2 until the cells are 50–80 % confluent.
3. Transfect the cells with 0.8 μg per dish of cDNA encoding phocus-2pp using LipofectAMINE 2000 according to the manufacturer's instructions (*see* **Note 1**).
4. Incubate the cells at 37 °C in 5 % CO_2 for 1–3 days after transfection.
5. Replace the medium with a serum-free Ham's F-12 medium supplemented with 0.2 % bovine serum albumin and incubate the cells for 2–6 h for serum starvation (*see* **Note 2**).
6. Wash with 1 mL of Hank's balanced salt solution (HBSS) and add 1 mL of HBSS for fluorescence imaging of the protein phosphorylation reaction by insulin receptor.

3.3 Fluorescence Imaging

1. Put the glass-bottomed dish on the stage of an inverted microscope equipped with a cooled CCD camera (*see* **Note 3**).
2. Image acquisition and processing are controlled by a PC connected to the CCD camera and a filter wheel using the MetaFluor software. A shutter for excitation in front of the xenon lamp is also controlled by the PC. Excitation light from

a 75 mW xenon lamp is passed through a 440 ± 10.5 nm band-pass filter for ECFP excitation (440AF21). The light is reflected onto the cells using a dichroic mirror (455DRLP). The emitted light is collected with a 40× or 63× objective and passed through a 480 ± 15 nm band-pass filter (480AF30) for ECFP emission and a 535 ± 13 nm band-pass filter (535AF26) for EYFP emission.

3. Define several factors for image acquisition, such as excitation power, time of exposure to the light, image acquisition interval, and binning (*see* **Note 4**).
4. By browsing the cells on the dish, choose moderately bright cells (*see* **Note 5**).
5. Draw a region of interest within the cell on the field of view of the CCD camera (*see* **Note 6**).
6. Start to acquire images every 2–10 s with the 440 ± 10.5 nm excitation. The software MetaFluor records fluorescence intensities of the ECFP and EYFP emissions of phocus-2pp, together with their emission ratios in the selected regions of the cells over time.
7. When the emission ratio of ECFP to EYFP reach a steady state, add 100 nM insulin to the dish to stimulate the cells, and monitor the protein phosphorylation reaction through the change in the emission ratio of phocus-2pp (Fig. 2).

3.4 Customizing the Biosensor

1. It should be mentioned that the present fluorescent indicator, phocus, has general applicability for a variety of protein kinases as well. Phocus has two main key components, the substrate domain and the phosphorylation recognition domain. By selecting these two components, phocus can be tailor-made for the protein kinases of interest. We would suggest to use substrate sequences that are selectively phosphorylated by the kinases of interest to rule out the possibility that the indicators are also phosphorylated by other kinases. We have developed fluorescent indicators for several key protein kinases, such as insulin receptor (receptor tyrosine kinase) [1], c-Src (non-receptor tyrosine kinase) [2, 3], Akt/PKB (serine/threonine protein kinases) [4], ERK (serine/threonine protein kinases) [5], JNK (serine/threonine protein kinases) [6] (Table 1).
2. The present indicators can be linked to subcellular localization sequences, such as nuclear localization sequence (NLS), nuclear export sequence (NES) and localization sequences to plasma membrane, Golgi membrane, and mitochondrial outer membrane, to visualize local activity of protein kinases in living cells (Table 1).

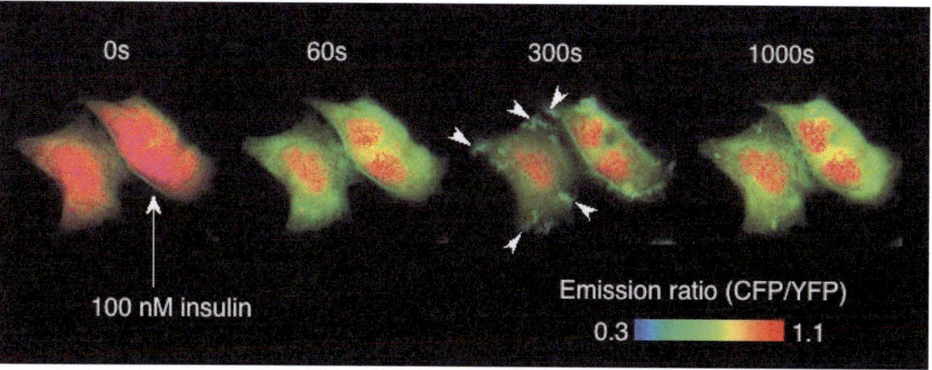

Fig. 2 Fluorescence imaging with phocus-2pp upon insulin stimulation. Pseudocolor images of the ECFP/EYFP emission ratio are shown before (time 0 s) and at 60, 300, and 1000 s after the addition of 100 nM insulin at 25 °C, obtained from the CHO-IR cells expressing phocus-2pp. Insulin-induced accumulation of phocus-2pp at the membrane ruffles is indicated by *white arrows* in the image at 300 s

4 Notes

1. The transfection mixture may be replaced with fresh 1.5 mL of Ham's F-12 medium supplemented with 10 % fetal bovine serum after 5–12 h of transfection, if desired.
2. This experimental step may not be necessary, depending upon cell types used for imaging.
3. Cells are viable and healthy for at least 60 min at ambient conditions. A CO_2 chamber and a temperature controller are required for long-term experiments.
4. If photobleaching of fluorescent proteins, in particular EYFP, is observed, use an ND filter in the range of 0.1–10 % to reduce excitation light intensity and/or decrease exposure time. Also, binning can sum the signal from multiple pixels on the CCD camera so that less light is required while keeping a good signal-to-noise ratio.
5. Avoid too bright cells to reproducibly obtain quantitative FRET signals. Also avoid too dim cells, because they often exhibit noisy signals.
6. The intensity in the region should be more than five times the background intensity.

Acknowledgements

This work was supported by grants from the Ministry of Education, Science and Culture, Japan and Japan Science and Technology Agency.

References

1. Sato M, Ozawa T, Inukai K, Asano T, Umezawa Y (2002) Fluorescent indicators for imaging protein phosphorylation in single living cells. Nat Biotechnol 20:287–294
2. Hitosugi T, Sasaki K, Sato M, Suzuki Y, Umezawa Y (2007) Epidermal growth factor directs sex-specific steroid signaling through Src activation. J Biol Chem 282:10694–10706
3. Hitosugi T, Sato M, Sasaki K, Umezawa Y (2007) Lipid raft-specific knockdown of Src family kinase activity inhibits cell adhesion and cell cycle progression of breast cancer cells. Cancer Res 67:8139–8148
4. Sasaki K, Sato M, Umezawa Y (2003) Fluorescent indicators for Akt/protein kinase B and dynamics of Akt activity visualized in living cells. J Biol Chem 278:30945–30951
5. Sato M, Kawai Y, Umezawa Y (2007) Genetically encoded fluorescent indicators to visualize protein phosphorylation by extracellular signal-regulated kinase (ERK) in single living cells. Anal Chem 79:2570–2575
6. Suzuki H, Sato M (2010) Genetically encoded fluorescent indicators to visualize protein phosphorylation by c-Jun NH_2-terminal kinase (JNK) in living cells. Supramol Chem 22:434–439

Chapter 12

Characterization of an Engineered Src Kinase to Study Src Signaling and Biology

Leanna R. Gentry, Andrei V. Karginov, Klaus M. Hahn, and Channing J. Der

Abstract

Pharmacologic inhibitors of protein kinases comprise the vast majority of approved signal transduction inhibitors for cancer treatment. An important facet of their clinical development is the identification of the key substrates critical for their driver role in cancer. One approach for substrate identification involves evaluating the phosphorylation events associated with stable expression of an activated protein kinase. Another involves genetic or pharmacologic inhibition of protein kinase expression or activity. However, both approaches are limited by the dynamic nature of signaling, complicating whether phosphorylation changes are primary or secondary activities of kinase function. We have developed rapamycin-regulated (RapR) protein kinases as molecular tools that allow for the study of spatiotemporal regulation of signaling. Here we describe the application of this technology to the Src tyrosine kinase and oncoprotein (RapR-Src). We describe how to achieve stable expression of this tool in cell lines and how to subsequently activate the tool and determine its function in signaling and morphology.

Key words FK506-binding protein, KRAS, mTOR, Oncogene, Pancreatic cancer, Rapamycin, Src, Tyrosine kinase

1 Introduction

Since Food and Drug Administration (FDA) approval of the first kinase inhibitor just over a decade ago [1], many small-molecule inhibitors of protein kinases have been developed successfully for cancer treatment. However, although the aberrant overexpression or activation of protein kinases have well-validated driver roles in cancer, the precise substrates critical for their cancer causing functions remain unclear. Establishing the role these kinase-dependent signaling events have in promoting cell transformation has been limited by the conventional methods utilized to study their function. A majority of the data implicating kinase activity in various disease phenotypes was collected using depletion of the kinase by RNA interference or pharmacologic inhibition, often resulting in

Hicham Zegzouti and Said A. Goueli (eds.), *Kinase Screening and Profiling: Methods and Protocols*, Methods in Molecular Biology, vol. 1360, DOI 10.1007/978-1-4939-3073-9_12, © Springer Science+Business Media New York 2016

off-target effects. Furthermore, these approaches provide an assessment of the loss of activity rather than the consequences of activation itself. For such studies, a common strategy is the expression of a constitutively active kinase. However, this approach is not applicable to many protein kinases. Furthermore, this approach allows for evaluation of cells stably overexpressing the protein kinase. All of these approaches make it difficult to identify transitional, directly kinase-activity-dependent steps leading to transformation.

A recently-developed tool circumvents many of these issues in the form of an inducible kinase [2]. This tool utilizes the well-characterized rapamycin-mediated heterodimerization of small FK506-binding protein (FKBP12) and an FKBP12-rapamycin-binding (FRB) domain of the mammalian target of rapamycin (mTOR). Rapamycin (MW 914.2 g/mol) is an FDA-approved drug used as an immunosuppressive and anticancer drug. This engineered protein contains a modified FKBP12 protein suitable for insertion into the middle of other proteins (insertable FKBP; iFKBP) (Fig. 1). The insertion of iFKBP into a structurally conserved portion of the kinase catalytic domain increases mobility

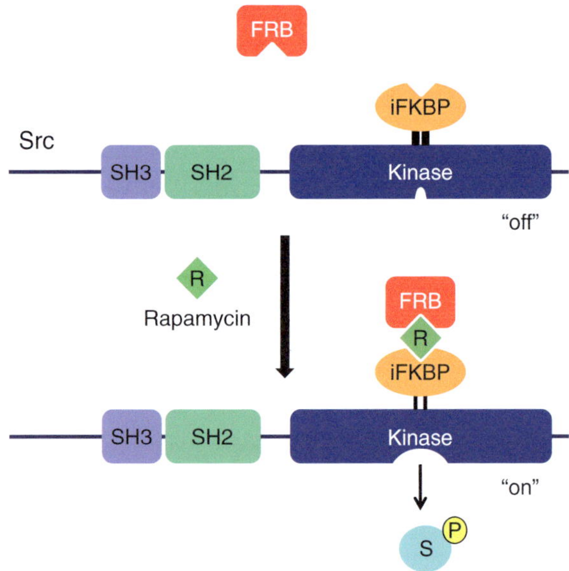

Fig. 1 RapR-Src construct: The Src tyrosine kinase is comprised of an N-terminal Srch homology domain 2 (SH2) followed by an SH3 domain, and a C-terminal kinase catalytic domain. A fragment of human FKBP12 (iFKBP12; residues 22–108) is inserted at a position in the kinase catalytic domain where it abrogates catalytic activity. A retrovirus vector encoding a Cerulean fusion with RapR-Src was generated. The FRB (FKBP12-rapamycin binding) domain of human mTOR (residues 2015–2114) was expressed as an mCherry-fusion protein. Binding to rapamycin (R) and FRB restores activity, leading to phosphorylation (P) of substrates (S)

of the region and thus decreases its ability to catalyze phosphate transfer. This construct is expressed in cells simultaneously with an exogenous FRB construct, where it remains inactive. Upon addition of rapamycin, the kinase domain is stabilized, restoring kinase activity and beginning downstream signaling. By controlling kinase activity through the addition of rapamycin or its non-immunosuppressive analogs, one can monitor events immediately after activation. Here, we discuss the application of this tool using the example of modified Src tyrosine kinase, termed RapR-Src (rapamycin-regulated Src). Src has long been studied due to its known contribution to progression of many cancers including colorectal, pancreatic, and prostate [3–5], and its early signaling events were of extreme interest when developing this tool. RapR-Src gives temporal control of Src activation. We recently demonstrated the usefulness of RapR constructs encoding different Src family members [6, 7]. Here we describe the implementation of RapR-Src to study Src signaling and biology in pancreatic cancer.

2 Materials

2.1 Expression Constructs and Plasmids

1. pBabe-Cerulean-RapR-Src: Retroviral expression vector (puromycin resistant) encodes an N-terminal cerulean fluorescent tag followed by full-length Src with the iFKBP in the kinase domain.
2. pLHCX-mCh-FRB retroviral expression vector (hygromycin resistant) encodes an N-terminal fluorescent tag fused to the FRB domain.
3. PCL10A-1 vector was used for retroviral packaging.
4. pBabe-tet-CMV-RapR-Src-GFP-Myc: Retroviral expression vector (puromycin resistant) encodes green fluorescent protein (GFP)-tagged RapR-Src under the control of a Tet-off expression system (Clontech).
5. pBabe-CMV-mCherry-FRB (puromycin resistant) retroviral expression vector encodes an mCherry-tagged FRB domain.

2.2 Cell Lines

1. HEK293T cells were obtained from the ATCC and maintained in Dulbecco's modified minimum essential medium (DMEM) supplemented with 10 % fetal calf serum (FCS).
2. Two immortalized human epithelial cell lines, HPDE and HPNE, and their matched mutant *KRAS* oncogene-transformed counterparts (HPDE-KRAS and HPNE-KRAS) cells were obtained from Michele Ouelette (University of Nebraska) and Ming Tsao (University of Toronto) [8, 9], respectively, and maintained in DMEM supplemented with 10 % FCS.

3. Mouse embryonic fibroblasts (Clontech Laboratories MEFs; Tet-off cell system).

2.3 Chemicals, Antibodies

1. Rapamycin: 1 mM stock solution made with ethanol (LC Laboratories). Store at −20 °C.
2. Polybrene: 8 mg/mL stock solution in water. Store at −20 °C.
3. Hygromycin: 50 mg/mL stock solution in sterile ddH_2O. Store at 4 °C.
4. Puromycin: 2 mg/mL stock solution in sterile ddH_2O. Store at −20 °C.
5. Anti-GFP mouse monoclonal antibody (clone JL-8; Clontech), 1 mg/mL.
6. Anti-Src rabbit monoclonal antibody (36D10; Cell Signaling).
7. Anti-FAK rabbit polyclonal antibody (Cell Signaling).
8. Anti-phospho-FAK (Tyr925) rabbit polyclonal antibody (Cell Signaling).
9. NP-40 lysis buffer: 1 % NP-40, 50 mM Tris pH 7.5, 10 mM $MgCl_2$, 150 mM NaCl, 5 % glycerol, 0.25 % Na-deoxycholate.
10. Phosphate-buffered saline (PBS): 137 mM NaCl, 2.7 mM KCl, 4.3 mM Na_2HPO_4, 1.47 mM KH_2PO_4, pH 7.4.
11. HEPES-buffered saline (HBS): 140 mM NaCl, 50 mM HEPES, 1.5 mM Na_2HPO_4. Adjust to pH 7.1 with 1 M NaOH, filter sterilize, and store at room temperature.
12. Calcium chloride (1.25 M): Dissolve $CaCl_2$ in distilled water and filter sterilize. Store at room temperature.
13. Poly-l-lysine: 0.01 % stock solution (Sigma), stored at 4 °C.

2.4 Imaging Components

1. #1.5 Glass cover slips (0.13–0.17 mm thick), 25 mm diameter. Store in 70 % ethanol.
2. Live cell imaging medium: L15 Leibovitz medium (Invitrogen) supplemented with 5 % FCS (*see* **Note 1**).
3. Mineral oil, sterile filtered, suitable for mouse embryo cell culture (Sigma-Aldrich).
4. Fibronectin: 1 mg/mL stock solution (BD Bioscience) dissolved in 0.5 M NaCl, 0.05 M Tris, pH 7.5 (*see* **Note 2**).
5. Collagen: Collagen Type I (rat tail) (BD Bioscience) suspended in 0.2 % acetic acid to 5 mg/mL.
6. Attofluor® cell chamber (Invitrogen) (*see* **Note 3**).
7. Cell culture plates; 35 mm.
8. Inverted microscope equipped with a 40× objective, CCD camera, a high-pressure mercury arc light source, and an open heated chamber (*see* **Note 4**).

3 Methods

3.1 Sequential Retroviral Infections of Pancreatic Epithelial Cells

1. Plate HEK293T cells at a density of 2.5×10^6 cells in a T25 flask 24 h prior to transfection.
2. To transfect cells, first add 5 μg of pLHCX-mCh-FRB DNA and 5 μg of pCL10A-1 DNA to 400 μL of HBS in a 1.5 mL microcentrifuge tube. Then add 100 μL of $CaCl_2$, briefly vortex the mixture, and add dropwise to the cells.
3. Five hours after transfection, change the growth medium.
4. The following day, change media 16 h prior to harvesting virus. On this day, split HPDE and HPDE-KRAS cells to a density of 20–30 % in T25 flasks.
5. To harvest virus, filter the conditioned medium from transfected 293T cell cultures using a syringe and a 0.45 μm filter. Add polybrene to the filtered medium to a concentration of 8 μg/mL.
6. Add 4 mL of growth medium back to flask containing to 293Ts to allow for a second transduction.
7. Add 2 mL of virus to target cells and incubate for 5 h, then replenish with fresh growth medium.
8. After 48 h, begin selecting for cells containing mCh-FRB using 50 μg/mL hygromycin. Selection should take 3–4 days (*see* **Note 5**).
9. After hygromycin selection, follow **steps 1–2** to infect cells containing mCh-FRB with pBabe-Cerulean-RapR-Src.
10. After 48 h, select cells using 1 μg/mL puromycin while maintaining a 25 μg/mL dose of hygromycin.
11. Continue to grow cells in the presence of 25 μg/mL of hygromycin and 0.5 μg/mL puromycin until you achieve 9×10^6 cells.
12. Sort cells positive for both Cerulean (excitation 433 nm, emission 475 nm) and mCherry (excitation 590 nm, emission 610 nm) fluorescence using cell sorter to enrich population of cells co-expressing RapR-Src-GFP-myc and mCherry-FRB (*see* **Note 6**).
13. Use untransduced cells and cells transduced with only one virus as controls for cell sorting. Continue growing sorted cells in the presence of 25 μg/mL of hygromycin and 0.5 μg/mL puromycin. If expression of RapR-Src is not desired while propagating the cells, then cells should be grown in the presence of doxycycline. Cell samples with different concentration of doxycycline should be tested to establish optimal concentration.

3.2 Simultaneous Retroviral Infection of Mouse Embryonic Fibroblasts

1. Prepare retroviruses as described in Subheading 3.1 using pBabe-tet-CMV-RapR-Src-GFP-myc and pBabe-CMV-mCherry-FRB constructs.
2. Plate MEFs to 2×10^5 cells onto 3 cm tissue culture plate the day before transduction (three plates total).

3. On the day of transduction remove media and add 1 mL of fresh media containing 16 μg/mL polybrene.
4. Add 2 mL 1:1 mixture of the two viruses to one plate and 1 mL of each virus to the remaining two dishes separately. Cells transduced with only one virus will be used as controls. Incubate overnight.
5. Change media and incubate for 48 h.
6. Re-plate cells into 6 cm plate and next day add puromycin at concentration 8 μg/mL to start selection.
7. Continue growing cells in the presence of puromycin until you achieve 3×10^6 viable cells.
8. Sort cells positive for both GFP (excitation 488 nm, emission 510 nm) and mCherry fluorescence using cell sorter to enrich population of cells co-expressing RapR-Src-GFP-myc and mCherry-FRB (*see* **Note 6**). Continue as described in Subheading 3.1, **steps 12** and **13**.

3.3 Activation of RapR System

1. The day before activation, plate HPDE or HPDE-KRas cells in 6-well plate.
2. To activate RapR-Src, add 0.5 μM rapamycin to growth medium, using 0.5 μM EtOH in a control well.
3. Three hours after activation, immediately wash cells with cold PBS and lyse cells on ice in ice-cold NP-40 lysis buffer. Incubate lysed cells in lysis buffer for 20 min on ice.
4. Centrifuge lysate for 15 min at $16{,}000 \times g$ to remove cell debris.
5. Take protein concentration by Bradford Assay and prepare 20 μg samples.
6. Run samples on SDS-PAGE and perform a western blot.
7. Use anti-Src, anti-FAK, anti-Y925 FAK, and anti-JL8 antibodies as directed by the manufacturer (Fig. 2).

3.4 Imaging

HPDE and HPDE-KRAS cells must be plated on collagen-coated cover slips for imaging.

1. Place one cover slip per well in a 6-well plate.
2. Wash cover slips with 70 % EtOH and aspirate.
3. Once cover slips are dried, coat with poly-l-lysine for 30 min in a sterile hood.
4. Aspirate and wash with sterile ddH_2O.
5. Next, cover cover slips with collagen at a concentration of 5 μg/mL in 0.2 % acetic acid and let sit overnight at 4 °C.
6. Aspirate solution and seed cells directly on collagen.
7. Plate HPDE or HPDE-KRAS cells at a density of 50,000 cells/well 24 h prior to imaging.

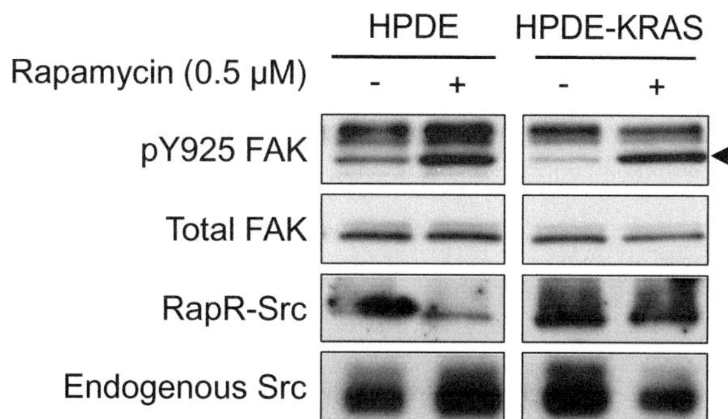

Fig. 2 RapR-Src activation induces phosphorylation of FAK tyrosine residue 925. HPDE and HPDE K-Ras cells co-expressing Cerulean-RapR-Src and mCherry-FRB were treated with 0.5 μM rapamycin to activate RapR-Src. Lysates were collected 3 h after treatment. Lysates were resolved by SDS-PAGE followed by blot analyses for phosphorylated Y925 FAK (indicated by *triangle*), total FAK, ectopically expressed RapR-Src (anti-JL8), and endogenous Src

8. On the day of imaging, aspirate media and wash cells one time with PBS. Place cover slip in an Attofluor® cell chamber and add 1 mL imaging media, ensuring that there are no leaks. Add 1 mL of mineral oil to top of media to reduce evaporation during imaging.
9. Place chamber on heated stage of microscope. Select cells that are expressing RapR-Src and FRB.
10. Image cells co-expressing Cerulean-RapR-Src and mCherry-FRB, taking images every min for 120 min. Mix 0.5 μL of 1 mM rapamycin solution with 100 μL of L15 Leibovitz Media and add to the cells (final concentration of 0.5 μM) 30 min after imaging has begun (*see* **Note 7**).
11. Continue imaging for the remaining 90 min. DIC imaging can be used to monitor cell movement and overall changes in cell morphology (i.e., protrusion formation and cell shape) (Fig. 3). Epifluorescence can be used to monitor RapR-Src, FRB, or any other fluorescently labeled co-transfected protein.

3.5 Imaging MEF Cells Expressing RapR-Src and FRB

1. Place a glass cover slip in 35 mm tissue culture plates or 6-well plates. Wash with 2–4 mL of PBS.
2. Incubate the cover slip in 2 mL of 5 μg/mL fibronectin solution in PBS at 37 °C overnight.
3. Wash the cover slip with PBS and add 2 mL of DMEM supplemented with 10 % FCS.
4. Plate transduced 50,000 MEF cells onto fibronectin-coated cover slip. Incubate in DMEM/10 % FCS medium for 2 h at 37 °C, 5 % CO_2 (*see* **Note 8**).
5. Preincubate mineral oil and imaging medium (L15 supplemented with 5 % FCS) at 37 °C for at least 30 min before imaging.

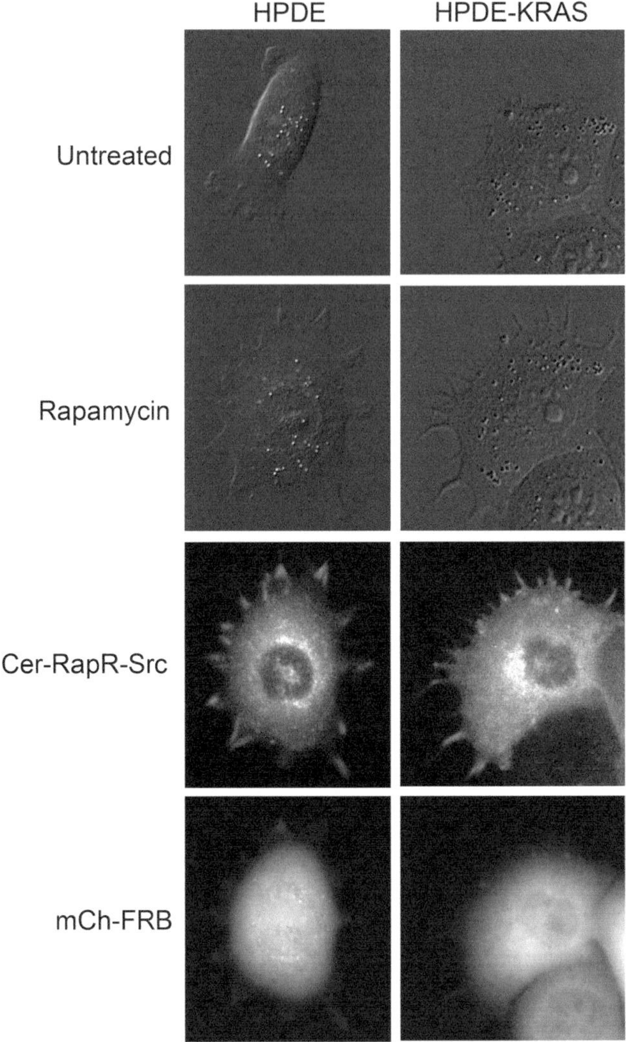

Fig. 3 RapR-Src activation induces cell spreading and protrusion formation. HPDE and HPDE-KRAS cells co-expressing Cerulean-RapR-Src and mCherry-FRB were treated with 0.5 μM rapamycin to activate RapR-Src. Image of rapamycin-treated cells was taken 1 h after treatment

6. Aspirate media and wash cells one time with PBS.
7. Place cover slip in an Attofluor® cell chamber and add 0.9 mL imaging media, ensuring that there are no leaks.
8. Add 1 mL of mineral oil to top of growth medium to reduce evaporation during imaging.
9. Place cell chamber onto heated stage of the microscope and select cells co-expressing RapR-Src-GFP and mCherry-FRB (*see* **Note 9**).

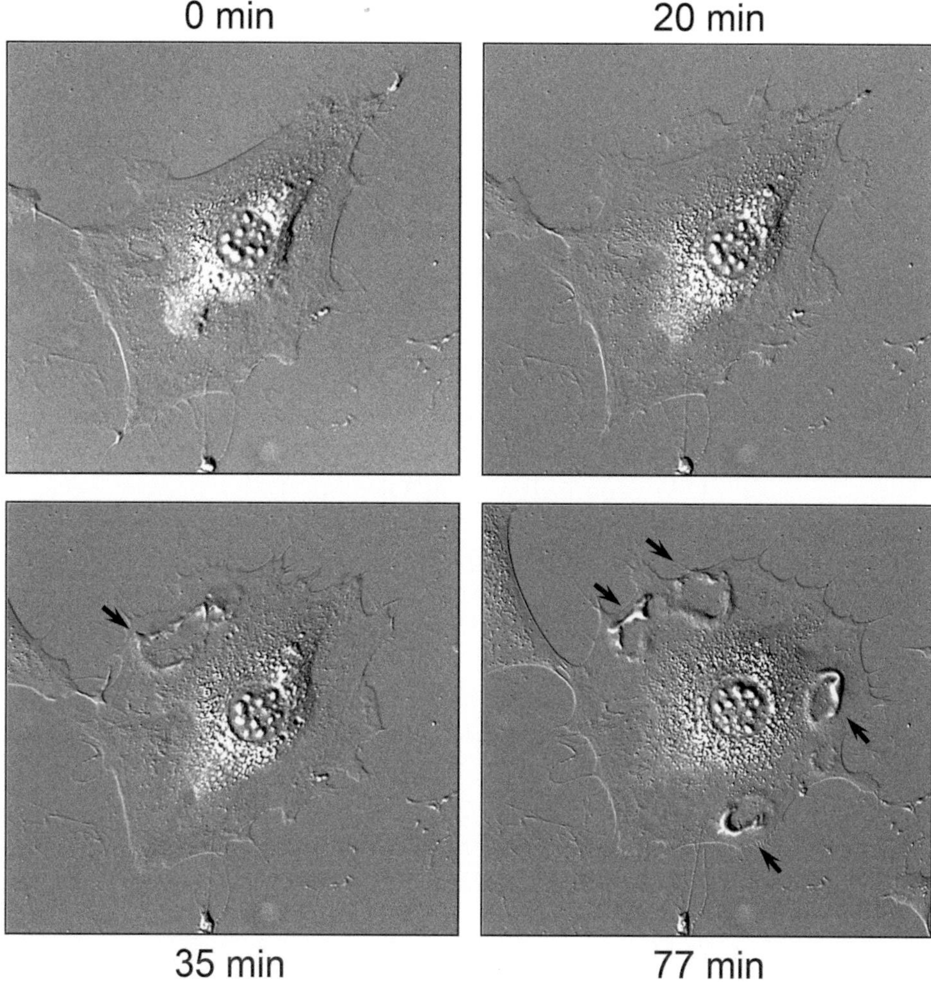

Fig. 4 RapR-Src activation induces cell spreading and dorsal wave formation (marked by arrows). MEFs co-expressing RapR-Src-GFP and mCherry-FRB were treated with 0.5 μM rapamycin to activate RapR-Src. Time after treatment is indicated

10. Image cells co-expressing RapR-Src-GFP and mCherry-FRB, taking images every min for 120 min. Mix 0.5 μL of 1 mM rapamycin solution with 100 μL of L15 Leibovitz Media and add to the cells (final concentration of 0.5 μM) 30 min after imaging has begun (*see* **Note 7**).
11. Continue imaging for the remaining 90 min. DIC imaging can be used to monitor cell movement and overall changes in cell morphology (i.e., protrusion formation and cell shape) (Fig. 4). Epifluorescence can be used to monitor RapR-Src, FRB, or any other fluorescently labeled co-transfected protein.

4 Notes

1. L15 medium is stored at 4 °C. FBS is stored separately at −20 °C. Fresh mix of L15 medium supplemented with FCS should be prepared on the day of the experiment. The amount depends on the number of imaging experiments performed; a minimum of 1 mL will be needed per experiment.
2. Fibronectin stock solution should be stored at 4 °C. Final solution used for coating cover slips should be prepared right before application.
3. Other devices suitable for live cell imaging using inverted epifluorescence microscopes can be used. The instrument should allow for addition of reagents during cell imaging.
4. We routinely use an Olympus IX-81 microscope equipped with a UPlanFLN 40× (Oil, NA 1.30) objective. All images are collected using a Photometrix CoolSnap ES2 CCD camera controlled by Metamorph software. Illumination for epifluorescence was provided from a high-pressure mercury arc light source.
5. Before selection, if expression of fluorescently tagged construct is not visible or is dim, perform a second round of viral transduction following **steps 2–4**.
6. We use Beckman Coulter MoFlo cell sorter equipped with tunable 355/405/568 nm laser and two fixed wavelength lasers (488 and 633 nm) provided by the UNC-Chapel Hill cell sorting facility.
7. Mix rapamycin with the media right before adding it to the cells. Make sure that you penetrate the oil layer when adding rapamycin solution to the cells.
8. It takes 1–2 h for MEFs to attach to the cover slips and spread.
9. A microscope equipped with a mechanized motorized stage enables consecutive imaging of several cells by selecting and logging the positions of cells expressing RapR-Src-GFP and mCherry-FRB. The number of positions depends on the time it takes all the images at one position and move to the next one, and on the desired imaging time intervals.

Acknowledgements

This work was supported, in whole or in part, by National Institutes of Health Grants CA042978 to C.J.D. and CA175747 to C. J. D. and K. M. H.

References

1. Druker BJ (2002) Perspectives on the development of a molecularly targeted agent. Cancer Cell 1:31–36
2. Karginov AV, Ding F, Kota P et al (2010) Engineered allosteric activation of kinases in living cells. Nat Biotechnol 28:743–747
3. Gargalionis AN, Karamouzis MV, Papavassiliou AG (2014) The molecular rationale of Src inhibition in colorectal carcinomas. Int J Cancer 134:2019–2029
4. Shields DJ, Murphy EA, Desgrosellier JS et al (2011) Oncogenic Ras/Src cooperativity in pancreatic neoplasia. Oncogene 30:2123–2134
5. Cai H, Babic I, Wei X et al (2011) Invasive prostate carcinoma driven by c-Src and androgen receptor synergy. Cancer Res 71:862–872
6. Chu PH, Tsygankov D, Berginski ME et al (2014) Engineered kinase activation reveals unique morphodynamic phenotypes and associated trafficking for Src family isoforms. Proc Natl Acad Sci U S A 111:12420–12425
7. Dagliyan O, Shirvanyants D, Karginov AV et al (2013) Rational design of a ligand-controlled protein conformational switch. Proc Natl Acad Sci U S A 110:6800–6804
8. Campbell PM, Groehler AL, Lee KM et al (2007) K-Ras promotes growth transformation and invasion of immortalized human pancreatic cells by Raf and phosphatidylinositol 3-kinase signaling. Cancer Res 67:2098–2106
9. Radulovich N, Qian JY, Tsao MS (2008) Human pancreatic duct epithelial cell model for KRAS transformation. Methods Enzymol 439:1–13

Chapter 13

Screening One-Bead-One-Compound Peptide Libraries for Optimal Kinase Substrates

Thi B. Trinh and Dehua Pei

Abstract

Protein kinases phosphorylate specific serine, threonine, and/or tyrosine residues in their target proteins, resulting in functional changes of the target proteins such as enzymatic activity, cellular location, or association with other proteins. For many kinases, their in vivo substrate specificity is at least partially defined by the amino acid sequence surrounding the phosphorylatable residue (or sequence specificity). We report here a robust, high-throughput method for profiling the sequence specificity of protein kinases. Up to 10^7 different peptides are rapidly synthesized on PEGA beads in the one-bead-one-compound format and subjected to kinase reaction in the presence of [γ-S]ATP. Positive beads displaying the optimal kinase substrates are identified by covalently labeling the thiophosphorylated peptides with a fluorescent dye via a disulfide exchange reaction. Finally, the most active hit(s) is identified by the partial Edman degradation-mass spectrometry (PED-MS) method. The ability of this method to provide individual sequences of the preferred substrates permits the identification of sequence contextual effects and non-permissive residues. This method is applicable to protein serine, threonine, and tyrosine kinases.

Key words Protein kinase, Substrate specificity, Sequence specificity, Peptide library, One-bead-one-compound library

1 Introduction

Approximately 30 % of human proteins are phosphorylated by 518 putative protein kinases on >100,000 serine, threonine, and tyrosine residues [1, 2]. In general, the phosphorylation events are highly specific with respect to both the kinase and the substrate protein. It is now established that protein kinases utilize a combination of several mechanisms to achieve exquisite substrate specificity in vivo, including temporal expression of the kinase and/or substrate, localization of the kinase and/or substrate to subcellular structures, protein-protein interaction through the use of recruiting domains/surfaces or scaffolding proteins, and interactions between the kinase active site and the linear sequence motif surrounding the phosphorylatable residue (or the intrinsic

Hicham Zegzouti and Said A. Goueli (eds.), *Kinase Screening and Profiling: Methods and Protocols*, Methods in Molecular Biology, vol. 1360, DOI 10.1007/978-1-4939-3073-9_13, © Springer Science+Business Media New York 2016

sequence specificity of the kinase domain) [3, 4]. For some protein kinases (e.g., protein kinase A), the intrinsic sequence specificity of the kinase active site is the major determinant of their in vivo substrate specificity, whereas other kinases show little sequence selectivity [3, 4].

One of the first tasks toward understanding the biological functions of a protein kinase is to identify its specific protein substrates. For kinases that recognize specific sequence motifs, a productive approach to substrate identification involves first defining the sequence specificity of the kinase domain using peptide substrates followed by searching the proteome against the consensus motif(s) [5–7]. Optimal peptide substrates also provide robust in vitro assays for the kinase as well as useful guide in designing specific inhibitors against the kinase. Several methods have previously been reported to determine the sequence specificity of protein kinases (reviewed in ref. [8]). However, the previous methods generally suffer from one or more drawbacks, e.g., inability to provide individual sequences and therefore detect any sequence contextual effect. Here we report a simple, robust, and general method for on-bead screening of one-bead-one-compound (OBOC) peptide libraries against a protein serine, threonine, or tyrosine kinase to determine its optimal peptide substrates.

For a protein kinase of unknown specificity, we recommend that one starts with a generic kinase substrate library in the form of X_5ZX_5NNBBRM-resin (library I), where B is β-alanine, Z is Ser, Thr, or Tyr, and X is any of the 19 proteinogenic amino acids except for methionine [replaced by L-norleucine (Nle or M)] and cysteine. The inclusion of a fixed Ser, Thr, or Tyr ensures that each peptide contains at least one phosphorylatable residue, although phosphorylation may also take place at any of the randomized positions. The linker sequence, NNBBRM, permits selective peptide release (cleavage after Met by CNBr) and facilitates peptide sequencing by partial Edman degradation-mass spectrometry (PED-MS; Arg provides a fixed positive charge and improves aqueous solubility) [9]. This library has a theoretical diversity of 19^9 or 2.6×10^{11} and should be synthesized on amino polyethylene glycol polyacrylamide (PEGA) resin (300–500 μm in water, ~1 million beads/g) (*see* **Note 1**). It is convenient to synthesize the library on ≤5 g of PEGA resin in a research lab setting (up to ~5 million different peptide sequences). Thus, the number of peptide sequences actually synthesized represents only a small percentage of the theoretical diversity. However, we have previously shown that the specificity profile of a protein can be unambiguously determined by sampling just a small fraction of the entire sequence space [8, 10–12]. If necessary, one may also synthesize and screen a secondary, biased library by fixing some of the random positions with preferred amino acids identified from the primary screening, in order to define the selectivity at less critical positions [8].

A major challenge associated with on-bead screening of enzymatic substrates is how to differentiate the reaction product(s) (typically <1 %) from a large excess of unreacted substrates. The key innovation of our method is a simple, robust assay for the kinase products (Fig. 1a). Briefly, the peptide library is treated for a limited amount of time with a kinase of interest in the presence of adenosine 5′-O-(3-thio)triphosphate ([γ-S]ATP) (instead of ATP), so that only beads carrying the optimal kinase substrates undergo a small amount of reaction (usually <1 %). The use of [γ-S]ATP results in the addition of a thiophosphoryl group to the positive beads, which are subsequently labeled with a fluorescent group (e.g., tetramethylrhodamine) through a disulfide exchange reaction. The fluorescent beads (Fig. 1b) can be manually isolated from the library with a micropipette under a fluorescence microscope and individually sequenced by PED-MS [9]. Substitution of [γ-S]ATP

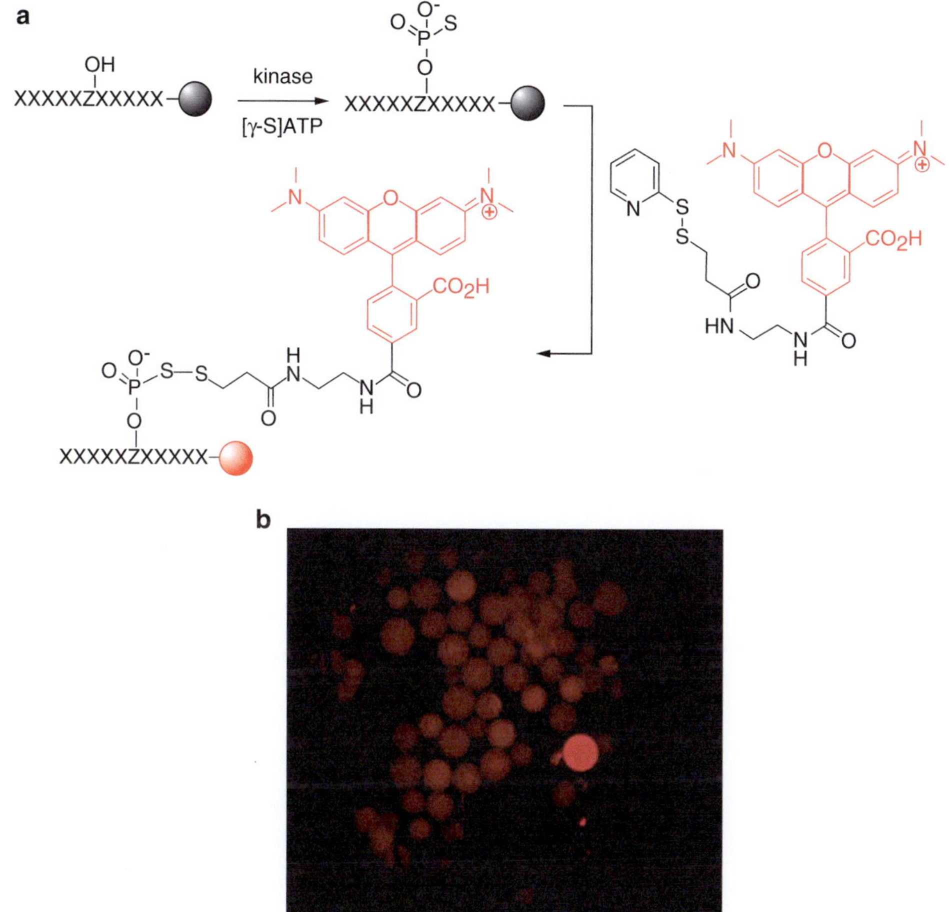

Fig. 1 Strategy for screening OBOC peptide libraries against protein kinases. (*a*) Reactions involved in the kinase screening strategy. X, random residues; Z, Ser, Thr, or Tyr. (*b*) A portion of the peptide library beads after the screening reactions (viewed under a fluorescence microscope)

for ATP decreases the catalytic activity of kinases by 15–30-fold but does not alter the sequence specificity [13, 14]. This screening method has a wide dynamic range and was able to profile the specificity of kinases that have catalytic efficiency (k_{cat}/K_M) ranging from 0.1 to 10^6 M^{-1} s^{-1} [8]. It provides individual peptide sequences, thus permitting the identification of not only amino acids that contribute positively to the kinase-substrate interaction (permissive residues), but also amino acids that negatively impact the kinase function (non-permissive residues), as well as any sequence covariance.

2 Materials

2.1 Library Synthesis

1. 0.2 mmol/g PEGA resin, 300–500 μm (Varian).
2. Fmoc-protected L-amino acids, 3-Fmoc-aminopropanoic acid (Fmoc-β-Ala-OH).
3. O-Benzotriazole-*N*,*N*,*N*′,*N*′-tetramethyluronium hexafluorophosphate (HBTU).
4. 1-Hydroxybenzotriazole hydrate (HOBt).
5. *N*-methylmorpholine (NMM).
6. 20 % Piperidine in DMF.
7. *N*,*N*-dimethylformamide (DMF).
8. Dichloromethane (DCM).
9. *N*-methylpyrrolidone (NMP).
10. CD_3CO_2D (Sigma-Aldrich).
11. $CH_3CD_2CO_2D$ (Sigma-Aldrich).
12. Trifluoroacetic acid (TFA).
13. Allyloxycarbonyl *N*-hydroxysuccinimide ester (Alloc-OSu).
14. Phenol, liquefied.
15. Double-distilled water (ddH_2O).
16. Thioanisole.
17. Ethanedithiol (EDT).
18. Triisopropylsilane (TIPS).
19. Anisole.
20. Kaiser test solution 1: 5 % (w/v) Ninhydrin in ethanol.
21. Kaiser test solution 2: 80 % (w/v) Phenol in ethanol.
22. Kaiser test solution 3: 2 % (v/v) of a 1 mM aqueous solution potassium cyanide, in pyridine.
23. Modified reagent K: 7.5 % phenol, 5 % water, 5 % thioanisole, 2.5 % EDT, 2.5 % TIPS, and 1.25 % anisole in TFA.

24. Rotary shaker.
25. Peptide synthesis vessels.
26. Spin columns, 1.2 mL bed volume (Bio-Rad).
27. Vacuum manifold with Luer connections.
28. Vacuum source.
29. Vortex mixer.

2.2 Synthesis of Labeling Reagent and Kinase Screening

1. 5-(and 6)-carboxytetramethylrhodamine, succinimidyl ester (TMR-NHS; Pierce).
2. Dimethyl sulfoxide, anhydrous (DMSO).
3. *N*-Boc-ethylenediamine.
4. *N*-Succinimidyl 3-[2-pyridyldithio]-propionate (SPDP).
5. Kinase buffer 1: 60 mM Tris, 150 mM NaCl, 5 mM $MnCl_2$, 2 mM dithiothreitol (DTT), 0.02 % Tween-20, pH 7.4.
6. Kinase buffer 2: 30 mM HEPES, 30 mM NaCl, 20 mM $MgCl_2$, 2 mM DTT, 0.02 % Tween 20, pH 7.4.
7. Adenosine 5′-O-(3-Thiotriphosphate), Tetralithium Salt ([γ-S]ATP; EMD Millipore).
8. 4 M Guanidine hydrochloride solution: Dissolve guanidine hydrochloride in water. Adjust pH to 7.4 with 6 M NaOH.
9. Labeling solution: 1:1 (v/v) NMP:50 mM HEPES, pH 7.4.
10. Selection solution: 50 mM NaCl, 0.02 % Tween-20.
11. Tris (2-carboxyethyl)phosphine hydrochloride (TCEP-HCl).
12. 50 mM TCEP solution: 28.6 mg TCEP-HCl in 2 mL selection solution.
13. Spin columns, 0.8 mL bed volume (Bio-Rad).
14. 35 mm petri dish.
15. Parafilm.
16. Orbital shaker.
17. Fluorescence microscope (Olympus).

2.3 Partial Edman Degradation and Mass Spectrometry for Sequencing Resin-Bound Peptides

1. Custom-designed reaction vessel (12 mm diameter, 20 mm height, a 10–20 μm glass frit, and 1 mm luer tip at the bottom (PED) vessel).
2. Pyridine.
3. TFA.
4. DCM.
5. Phenylisothiocyanate, liquid (PITC; Sigma-Aldrich).
6. *N*-(9-Fluorenylmethoxycarbonyloxy)succinimide (Fmoc-OSu).

7. Triethylamine (TEA).
8. Dimethyl sulfide.
9. Ammonium iodide.
10. Cyanogen bromide, 40 mg/mL solution in 70 % TFA in water.
11. Acetonitrile.
12. 4-Hydroxy-α-cyanocinnamic acid (α-CCA).
13. α-CCA solution: 10 mg/mL α-CCA in 50:50:0.1 (v/v) acetonitrile/water/TFA.
14. Tetrakis(triphenylphosphine)palladium(0) [$Pd(PPh_3)_4$].
15. Triphenylphosphine (PPh_3).
16. Tetrahydrofuran, anhydrous (THF).
17. *N*-methylaniline.
18. Sodium dimethyldithiocarbamate (SDDC).
19. Dissection microscope (10× to 40× magnification).
20. SpeedVac concentrator (Thermo Scientific).
21. Matrix-assisted laser desorption ionization (MALDI) sample plate (Bruker).
22. MALDI-TOF system (Bruker).

3 Methods

3.1 Synthesis of Solid-Phase Peptide Library

3.1.1 Synthesis of Linker Region: NNBBRM-Resin

1. Weigh about 2 g of PEGA resin (~20 g wet resin) into a 30 mL glass or polypropylene reaction vessel, equipped with a filter. The diameter of the reaction vessel should be sufficient so that the height of the resin bed is less than 2 in. PEGA resin typically comes as soaked in ethanol at 10 % concentration.
2. Wash the resin five times with DMF. This is typically done by covering the resin completely with DMF, and then drain.
3. Swell the resin by incubating in DMF for 20 min on a rotary shaker.
4. Drain the solvent. The resin is now ready for library synthesis.
5. Dissolve 4 eq. of Fmoc-Met-OH (594.2 mg), 4 eq. of HBTU (606.8 mg), and 4 eq. of HOBt (244.8 mg) in 20 mL of DMF. Add 8 eq. of NMM (352 μL), mix well for 10 s and add to resin. Incubate the mixture on a rotary shaker for 90 min.
6. Wash the resin with DMF (3 times), and then with DCM (3 times).
7. Prepare Kaiser test mixture by adding 40 μL of each Kaiser test solution (1, 2, and 3) into a glass test tube. A control tube should also be prepared with the same amount of each reagent.

8. Using a pipet tip, add a small amount of resin (~100 beads) to the test tube. No resin is added to the control tube.
9. Heat the mixtures at 100 °C for 3 min.
10. Examine the color of each mixture. A yellow to brown color indicates a negative test (no free $-NH_2$ is present, coupling is complete). The color of sample tube should be very similar, if not identical, to the color of the control tube. (If the coupling is incomplete, *see* **Note 2**.)
11. Remove the N-terminal Fmoc group by treating the resin with 20 % piperidine in DMF for 20 min.
12. Wash the resin with DMF (five times), DCM (four times).
13. Perform a Kaiser test to ensure the 20 % piperidine in DMF solution is functioning properly. A dark purple solution after heating should be observed (presence of a free $-NH_2$ group).
14. Repeat **steps 5–12** for the rest of the linker positions.

3.1.2 Synthesis of Random Positions: X_5ZX_5-Linker by Split-and-Pool Synthesis

1. After the removal of the N-terminal Fmoc group of the linker sequence, the resin is now ready to be split (*see* **Note 3**).
2. Suspend the resin in 20 mL of 1:1 DCM/DMF mixture in a reaction vessel. The resin density should be uniform throughout the solvent. The reaction vessel can be inverted regularly to ensure uniformity of resin density.
3. Split the resin into 19 (for Tyr library) or 20 (for Ser/Thr library) 2 mL reaction vessels using a 1000 μL pipet. The volume of each split should be around 800 μL.
4. Add 1:1 DCM/DMF mixture to the remaining resin and repeat the splitting until no resin is left in the 30 mL reaction vessel (*see* **Note 4**).
5. Each 2 mL reaction vessel should be clearly labeled with the name of an amino acid.
6. Treat the resin with 4 eq. of Fmoc-AA-OH, HBTU, HOBt and 8 eq. of NMM for 90 min, where AA is the appropriate amino acid (*see* **Note 5**).
7. Wash the resin with DMF (3 times).
8. Treat the resin with 4 eq. of Fmoc-AA-OH, HBTU, and HOBt and 8 eq. of NMM for 60 min, where AA is the appropriate amino acid.
9. Wash the resin with DMF (three times), and DCM (three times).
10. Perform Kaiser test on three representative reactions. Ile is usually chosen due to the low coupling efficiency. The other two are randomly chosen. A control test (no resin) should also be performed. Proceed if all three Kaiser tests are negative.

11. Suspend the resin in 1:1 DCM/DMF mixture.
12. Transfer all the resin as slurry into the 30 mL reaction vessel using a 1000 μL pipet. Repeat the transfer until no resin is left in each 2 mL vessel.
13. Treat the resin with 20 % piperidine in DMF for 20 min.
14. Wash the resin with DMF (5 times) and DCM (4 times).
15. Repeat **steps 2–14** for four positions.
16. For coupling of fixed amino acid, resin was treated with 4 eq. of HBTU, and HOBt; 4 eq. of appropriate Fmoc-protected amino acid; and 8 eq. of NMM in DMF for 90 min. For Tyr library, 2 g of resin was treated with 4 eq. of Fmoc-Tyr-OH. For Ser/Thr library, resin was split into two equal portions; each was coupled with 4 eq. of Fmoc-Ser-OH or Fmoc-Thr-OH.
17. Wash the resin with DMF (three times) and DCM (three times).
18. Perform Kaiser test to ensure that the coupling is complete.
19. Perform split and pool synthesis (**steps 2–14**) for the 5N-terminal random positions.
20. After library synthesis and removal of N-terminal Fmoc group, the library is treated with 15 eq. of Alloc-OSu and 5 eq. of NMM in DMF for 30 min.
21. Wash the library with DMF (five times) and DCM (ten times).
22. Side-chain deprotection was carried out by treating the resin with 20 mL of modified reagent K for 2 h.
23. Wash the resin with TFA (three times), DCM (ten times), and DMF (ten times).
24. Suspend the resin in DMF and store at −20 °C. The library is now ready for screening.

3.2 Screening the Library Against Kinase

3.2.1 Synthesis of Fluorescence Label

1. Dissolve TMR-NHS in DMF at 40 mM concentration (10.5 mg in 0.5 mL).
2. Add 1.1 eq. of N-Boc-ethylenediamine (3.3 μL).
3. Stir the reaction overnight in the dark.
4. Remove DMF by SpeedVac concentrator.
5. Add 1 mL of toluene and mix well.
6. Remove the solvent by SpeedVac concentrator.
7. Dissolve the crude product (TMR-NH_2-Boc) in 1 mL of TFA.
8. Stir the reaction for 2 h in the dark.
9. Remove TFA by SpeedVac.
10. Add 1 mL of toluene and remove the solvent by SpeedVac.
11. Dissolve the crude product (TMR-NH_2) in 400 μL of anhydrous DMSO (50 mM concentration).

12. Weigh out 1.2 mg of SPDP into a 1.5 mL microcentrifuge tube.
13. Add 78.5 μL of TMR-NH_2 solution.
14. Add 0.55 μL of triethylamine.
15. Incubate for 2 h in the dark.
16. Store the crude product (TMR-S-S-Py) in DMSO at −20 °C.

3.2.2 Library Screening

1. Transfer 100 mg of resin (as slurry in DMF) into a 2 mL polypropylene spin column.
2. Wash with DMF (three times), ddH_2O (five times), and kinase buffer (five times).
3. Transfer the resin into a 1.5 mL microcentrifuge tube as slurry in kinase buffer.
4. The resin should settle at the bottom of the tube, leaving a clear supernatant. Carefully pipet and discard the supernatant, leaving the resin in minimal (~290 μL) amount of kinase buffer (*see* **Note 6**).
5. Prepare 100 mM solution of [γ-S]ATP in 50 mM HEPES (pH 7.4). Dissolve 2 mg of [γ-S]ATP in 36 μL of 50 mM HEPES. Mix well with a pipette.
6. Add 6 μL of 100 mM [γ-S]ATP solution to the resin.
7. Add the kinase to the reaction mixture. The amount of kinase (and incubation time) depends on the enzymatic activity. Usually a final concentration of 5 μM of kinase is recommended for initial screening.
8. Invert the microcentrifuge tube and lightly tap the side, so the whole reaction mixture settles in the cap and top of the tube.
9. Wrap the cap of the tube with parafilm.
10. Fix the microcentrifuge tube (in the inverted position) onto an orbital shaker using masking tape.
11. Mix the reaction at 200 rpm, 30 °C for 3–40 h, depending on the enzymatic activity of the kinase.
12. Invert the tube back to its upright position and lightly tap the side so the reaction mixture settles at the bottom.
13. Transfer the reaction mixture to a 2 mL spin column.
14. Wash the resin with kinase buffer (five times), ddH_2O (five times) and 4 M guanidine-HCl, pH 7 (three times).
15. Incubate the resin in 4 M guanidine-HCl three times, 15 min each time on a rotary shaker.
16. Wash the resin with water (ten times) and labeling solution (five times).
17. Prepare 100 μM solution of TMR-NH_2 in DMSO.

18. Add 2 μL of TMR-NH_2 solution to 1.6 mL of labeling solution (0.125 μM) and mix well.
19. Add this mixture to the resin and incubate for 2 h in the dark.
20. Wash the resin with labeling solution (five times), DMF (five times), and selection solution (five times).
21. Transfer the resin to a petri dish using selection solution.
22. Visualize the beads under a fluorescence microscope, isolate, and discard all fluorescent beads using a micropipette (*see* **Note 7**).
23. Transfer all the non-fluorescent beads into a 2 mL spin column.
24. Prepare 100 μM solution of TMR-S-S-Py in DMSO.
25. Add 2 μL of TMR-S-S-Py solution to 1.6 mL of labeling solution (0.125 μM) and mix well.
26. Add this mixture to the resin and incubate for 2 h in the dark.
27. Wash the resin with labeling solution (five times), DMF (five times), and selection solution (five times).
28. Transfer the resin to a petri dish using selection solution.
29. Visualize the beads under a fluorescence microscope and isolated the fluorescent beads.
30. Prepare 50 mM TCEP solution. Transfer TCEP solution to a petri dish.
31. Transfer fluorescent beads to the petri dish containing TCEP solution and visualize the disappearance of fluorescence signal. Positive beads should lose their fluorescence upon TCEP treatment.
32. Transfer the positive beads (now non-fluorescent) to a microcentrifuge tube and store at 4 °C.

3.3 Sequencing of Hit Peptides by PED-MS

3.3.1 Removal of N-Terminal Alloc Group

1. Add 500 μL of DMF to the microcentrifuge tube containing hits from the screening.
2. Transfer the mixture to a PED vessel.
3. Add 500 μL of DCM to the microcentrifuge tube.
4. Pour the mixture into the PED vessel. All beads should now be in the PED vessel.
5. Wash the beads with DMF (three times) and THF (three times).
6. Treat the beads with 5 mg of $Pd(PPh_3)_4$, 50 mg of PPh_3, and 40 μL of N-methylaniline in 1 mL of THF for 2 h.
7. Wash the beads with THF (five times) and DMF (five times).
8. Wash the beads with 1 mL of 1 % (w/v) SDDC in DMF.
9. Incubate the beads with 1 mL of 1 % SDDC in DMF for 15 min.
10. Wash the beads with DMF (five times) and pyridine (five times). The peptides are now ready for sequencing by PED/MS.

3.3.2 PED

1. Prepare 2:1 (v/v) pyridine/water mixture containing 0.1 % triethylamine (mix 8 mL of pyridine, 4 mL of water and 12 μL of triethylamine).
2. Prepare 7.8 mM solution of Fmoc-OSu in pyridine.
3. Mix 160 μL of Fmoc-OSu solution with 148 μL of 2:1 pyridine/water mixture.
4. Add 12 μL of PITC and mix briefly (10 s).
5. Add this mixture (320 μL) to the PED vessel.
6. Allow the reaction to proceed for 6 min and then wash with pyridine (two times) and DCM (three times).
7. Wash the beads with TFA (one time) and incubate in 1 mL of TFA for 6 min.
8. Drain the solvent, add 1 mL of TFA, and incubate for 6 min.
9. Drain the solvent and wash the beads with DCM (three times) and pyridine (two times).
10. Repeat **steps 3–9** for the desired number of cycles.
11. After the second TFA treatment of the last PED cycle, suspend the beads in 500 μL of TFA.
12. Add 30 μL of dimethyl sulfide to the reaction.
13. Add 500 μL of TFA and 30 mg of ammonium iodide and let the reaction proceed for 20 min.
14. Wash the beads with TFA (two times), water (ten times), DMF (three times), and 20 % piperidine in DMF (one time).
15. Treat the beads with 1 mL of 20 % piperidine in DMF for 20 min.
16. Wash the beads with DMF (three times) and water (five times).
17. Suspend the beads in 1 mL of water and transfer them to a petri dish using a pipette. Repeat this transfer until all beads are present in the petri dish (*see* **Note 8**).
18. Transfer the beads into individual 1.5 mL microcentrifuge tubes (one bead/tube).
19. Add 20 μL of 40 mg/mL CNBr in 70 % TFA in water to each tube and allow the reaction to proceed in the dark for 12–14 h.
20. Dry the samples in SpeedVac and store at 4 °C.

3.3.3 MALDI-TOF MS Analysis

1. Add 5 μL of 50:50:0.1 (v/v) acetonitrile/water/TFA to each sample tube.
2. Vortex the tubes for 10 s and centrifuge at 870 × *g* for 10 s.
3. Repeat **step 2** twice.
4. Mix 1 μL of this solution with 2 μL of α-CCA solution.
5. Spot 1 μL of the mixture onto a MALDI sample plate.

6. Perform mass spectrometric analysis on a Bruker Microflex MALDI-TOF instrument (or any other MALDI-TOF model).
7. Mass spectra can be analyzed with Moverz or FlexAnalysis software.

4 Notes

1. The choice of PEGA or other resins that are permeable to relatively large proteins is crucial for successful on-bead enzymatic reactions. Popular resins such as TentaGel are not permeable to macromolecules and should be avoided.
2. Repeat any incomplete coupling reaction with 4 eq. Fmoc-amino acid, 4 eq. of HATU, and 8 eq. of NMM.
3. It is not recommended to dry PEGA resin and therefore resin splitting is best carried out by pipetting the resin slurry suspended in solvents.
4. Despite uniformity during the splitting, the first few vessels usually get slightly more resin than the subsequent ones. The second round of splitting should be done in reverse order (transfer resin to last vessel first) to achieve an equal amount of resin in each vessel.
5. 3.5 % (mol/mol) of CD_3CO_2D is added to the coupling solution of Fmoc-Leu-OH and Fmoc-Lys-OH, while 3.5 % (mol/mol) of $CH_3CD_2CO_2D$ is added to the solution of Fmoc-Nle-OH. This allows for the differentiation of isobaric amino acids during MS analysis [9].
6. The resin itself has substantial volume. It is important that the total reaction volume be large enough to ensure proper mixing of the reaction contents and yet minimal in order to conserve enzyme and reagents. A final volume of 300 μL works well for 100 mg of resin.
7. The number of false-positive beads, caused by binding of the fluorescent dye directly to the peptides, was typically 10–30 for 100 mg of resin.
8. Alternatively, hold the PED vessel upside down above a petri dish and spray it with a water bottle to wash the beads out of the vessel and into the petri dish.

Acknowledgements

The work in this laboratory is supported by the National Institutes of Health (GM062820 and CA0132855).

References

1. Hornbeck PV, Kornhauser JM, Tkachev S, Zhang B, Skrzypek E, Murray B, Latham V, Sullivan M (2012) PhosphoSitePlus: a comprehensive resource for investigating the structure and function of experimentally determined post-translational modifications in man and mouse. Nucleic Acids Res 40:261–270
2. Manning G, Whyte DB, Martinez R, Hunter T, Sudarsanam S (2002) The protein kinase complement of the human genome. Science 298:1912–1934
3. Ubersax JA, Ferrell JE Jr (2007) Mechanisms of specificity in protein phosphorylation. Nat Rev Mol Cell Biol 8:530–541
4. Zhu G, Liu Y, Shaw S (2005) Protein kinase specificity. A strategic collaboration between kinase peptide specificity and substrate recruitment. Cell Cycle 4:52–56
5. Hutti JE, Jarrell ET, Chang JD, Abbott DW, Storz P, Toker A, Cantley LC, Turk BE (2004) A rapid method for determining protein kinase phosphorylation specificity. Nat Methods 1:27–29
6. Fujii K, Zhu G, Liu Y, Hallam J, Chen L, Herrero J, Shaw S (2004) Kinase peptide specificity: improved determination and relevance to protein phosphorylation. Proc Natl Acad Sci U S A 101:13744–13749
7. Mok J, Kim PM, Lam HY, Piccirillo S, Zhou X, Jeschke GR, Sheridan DL, Parker SA, Desai V, Jwa M, Cameroni E, Niu H, Good M, Remenyi A, Ma JL, Sheu YJ, Sassi HE, Sopko R, Chan CS, De Virgilio C, Hollingsworth NM, Lim WA, Stern DF, Stillman B, Andrews BJ, Gerstein MB, Snyder M, Turk BE (2010) Deciphering protein kinase specificity through large-scale analysis of yeast phosphorylation site motifs. Sci Signal 3:ra12
8. Trinh TB, Xiao Q, Pei D (2013) Profiling the substrate specificity of protein kinases by on-bead screening of peptide libraries. Biochemistry 52:5645–5655
9. Thakkar A, Wavreille AS, Pei D (2006) Traceless capping agent for peptide sequencing by partial Edman degradation and mass spectrometry. Anal Chem 78:5935–5939
10. Sweeney MC, Wavreille A, Park J, Butchar JP, Tridandapani S, Pei D (2005) Decoding protein-protein interactions through combinatorial chemistry: sequence specificity of SHP-1, SHP-2, and SHIP SH2 domains. Biochemistry 44:14932–14947
11. Luechapanichkul R, Chen X, Taha HA, Vyas S, Guan X, Freitas MA, Hadad CM, Pei D (2013) Specificity profiling of dual specificity phosphatase vaccinia VH1-related (VHR) reveals two distinct substrate binding modes. J Biol Chem 288:6498–6510
12. Chen X, Ren L, Kim S, Carpino N, Daniel JL, Kunapuli SP, Tsygankov AY, Pei D (2010) Determination of the substrate specificity of protein-tyrosine phosphatase TULA-2 and identification of Syk as a TULA-2 substrate. J Biol Chem 285:31268–31276
13. Allen JJ, Li M, Brinkworth CS, Paulson JL, Wang D, Hübner A, Chou WH, Davis RJ, Burlingame AL, Messing RO, Katayama CD, Hedrick SM, Shokat KM (2007) A semisynthetic epitope for kinase substrates. Nat Methods 4:511–516
14. Xu WQ, Zhou SY, Eck MJ, Cole PA (1998) Peptide and protein phosphorylation by protein tyrosine kinase Csk: insights into specificity and mechanism. Biochemistry 37: 165–172

Chapter 14

Determination of the Substrate Specificity of Protein Kinases with Peptide Micro- and Macroarrays

Shenshen Lai, Dirk F.H. Winkler, Hong Zhang, and Steven Pelech

Abstract

Elucidation of the key determinants for the phosphorylation site specificities of protein kinases facilitates identification of their physiological substrates, and serves to better define their critical roles in the signaling networks that underlie a multitude of cellular activities. Albeit with some apparent limitations, such as the lack of contextual information for secondary substrate-binding sites, the synthetic peptide-based approach has been adopted widely for the kinase specificity profiling studies, especially when they are used in an array format, which permits the screening of large numbers of potential peptide substrates in parallel. In this chapter, we present detailed protocols for determining protein kinase substrate specificity using an approach that involves both peptide microarrays and macroarrays. In particular, SPOT synthesis on macroarrays can be used to follow up on in silico predictions of protein kinase substrate specificity with predictive algorithms.

Key words Protein kinase specificity, Phosphorylation, SPOT synthesis, Peptide array, Peptide microarray, Substrate screening

1 Introduction

Protein phosphorylation catalyzed by protein kinases is the most common posttranslational mechanism to transduce extracellular signals intracellularly in eukaryotic organisms. In humans, about 2 % of the protein-coding genes specify at least 536 known protein kinases [1], which may phosphorylate as many as one million phosphosites in the phosphoproteome, with over 200,000 phosphosites already confirmed experimentally (www.phosphonet.ca). Despite this apparent promiscuity amongst protein kinases, they still demonstrate rather restricted physiological substrate specificity such that signaling networks typically operate with high fidelity and exquisite control. Such selective activation/deactivation of a subset of intracellular signaling pathways mediated by protein kinases relies heavily on the ability of these enzymes to differentiate their *bona fide* substrates *in vivo*. Indeed, dysregulation of protein

Hicham Zegzouti and Said A. Goueli (eds.), *Kinase Screening and Profiling: Methods and Protocols*, Methods in Molecular Biology, vol. 1360, DOI 10.1007/978-1-4939-3073-9_14, © Springer Science+Business Media New York 2016

phosphorylation events is implicated in over 400 types of human diseases, including cancer, diabetes, cardiovascular, neurological, and immunological disorders. Mutations in protein kinase genes may affect their enzymatic activity, substrate specificity, location, and stability. As the genomes of more individuals become sequenced for diagnostic purposes, prediction of changes in these parameters will first require better knowledge of the subtle differences between protein kinases in these properties.

Several molecular mechanisms contribute to protein kinase-substrate pairing specificity, such as the co-expression, co-localization, and co-activation of the kinase and its substrates, and protein-protein interactions mediated through direct docking sites or indirectly mediated via adaptor/scaffolding proteins. However, the selectivity of a particular protein kinase for its substrates is influenced to a high degree by molecular recognition of the amino acid sequences surrounding the phosphorylation sites [2–6]. Exploring the specificity of protein kinases using synthetic peptides has been a fruitful endeavor in cell signaling research for half a century [7–9]. Still, only a relatively small fraction of the human protein kinases have been systematically and comprehensively studied for their substrate selectivity, leaving lots of "orphans" with unknown connections to the rest of the cell signaling apparatus.

Simple alignment of the amino acid residues in phosphorylation sites of known substrate proteins and synthetic peptide substrates of many protein kinases provides insights into the amino acid preferences of individual protein kinases. This defines what is termed as a consensus substrate recognition sequence for a protein kinase. Favored amino acid residues in substrates show up at the highest frequencies at particular positions surrounding the phospho-acceptor amino acid.

The vast majority of the eukaryotic protein kinases, referred to as "typical," feature a homologous catalytic domain with approximately 250 amino acid residues, including about 16 in highly conserved subdomains. These kinase subdomains are particularly critical for binding the ATP substrate and catalytic activity. From careful alignment of the consensus recognition sequences of protein kinases with their catalytic domain sequences, it has been feasible to identify the substrate determining residues of typical kinases. To predict the consensus recognition sequences for 492 human protein kinases domains, we have developed algorithms that have been trained with empirical data for over 14,000 known kinase-protein substrate pairs and 8000 kinase-peptide substrate pairs [10]. The results from these *in silico* analyses have been posted with open access on our PhosphoNET website (www.phosphonet.ca). This data is an excellent starting place for testing peptide substrates for target protein kinases of interest that have not been previously well studied.

In recent years, development of high-throughput techniques has made it feasible to screen for the specificity of a protein kinase of interest against a large set of peptides spotted on cellulose membrane or microarray surfaces [11–14]. The SPOT technique allows flexible and inexpensive synthesis of large number of peptides on cellulose membranes [15–20]. These peptides can be released and immobilized on microarray slides, or used directly for phosphotransferase assays as peptide macroarrays. Different strategies can be taken to select a library of peptides for kinase profiling [21–23]. A knowledge-based peptide library of confirmed physiological phosphorylation sites can be chosen in a microarray screening, while sequences modified from known protein substrates are also applicable. Quantitative analysis of the microarray image permits derivation of a consensus substrate sequence, which can later be used as a wild-type template of a substitutive library to be tested on the macroarray. In other cases, combinatorial libraries are employed for de novo screenings on the SPOT membrane [24].

In this chapter, we describe the procedures of kinase substrate determination with peptide macro- and microarrays. While they have been demonstrated to very successfully identify an overall preference of amino acids at a given position, one limitation of this technique is the potential loss of contextual information unique to proteins. This is, however, overweighed by the advantages of their flexibility, easiness, convenience, and ever-decreasing costs. We recommend proceeding from an *in silico* analysis such as offered by PhosphoNET or a high-content peptide microarray to define the promising candidate peptide substrates. Systematic replacements of the amino acids in these peptides by SPOT synthesis on macroarrays can then be used to further improve specific kinase recognition. A specific substrate peptide identified using the above methods can serve as a powerful investigative tool to assay protein kinase activity *in vitro* with high sensitivity and selectivity, and provide insight into the regulation of a target protein kinase in cell signaling networks.

2 Materials

2.1 SPOT Synthesis and Preparation of Peptide Microarrays (See Note 1*)*

1. Solvents: *N,N'*-dimethylformamide (DMF), methanol (MeOH), ethanol (EtOH), *N*-methylpyrrolidone (NMP), diethylether (DEE), dichloromethane (methylene chloride, DCM) (*see* **Note 2**).
2. Membranes for the peptide synthesis are prepared from filter paper Whatman 540 or Whatman 50 (Whatman) [18, 25].
3. Diisopropylcarbodiimide (DIPC, DIC; Fluka), *N-Methylimidazole* (NMI; Sigma), and Fmoc-β-alanine, Fmoc-glycine (GLS Biochem) (*see* **Note 3**).

4. Amino functionalization solution: For 12.5 ml, 782 mg Fmoc-β-alanine or 750 mg Fmoc-glycine prepared in DMF. Add 468 μl DIC and 396 μl NMI.
5. Staining solution: 0.002 % (w/v) bromophenol blue (BPB; Sigma) in MeOH.
6. Pentafluorophenyl ester-preactivated amino acid derivatives with protecting groups according to the Fmoc strategy (OPfp esters; EMD Millipore, GLS Biochem, and Bachem) [26, 27] (*see* **Note 4**).
7. Piperidine solution: 20 % Piperidine (Sigma) in DMF.
8. Capping solution: 2 % (v/v) acetic anhydride (Sigma) and 2 % (v/v) ethyl-diisopropylamine (DIPEA, DIEA; Sigma) in DMF.
9. Deprotection solution A: Trifluoroacetic acid (TFA) containing 4 % (v/v) DCM, 3 % (v/v) triisopropylsilane or triisobutylsilane (TIPS or TIBS; Fluka), 2 % (v/v) distilled water, 1 % (w/v) thioanisole (methyl phenyl sulfide; Alfa Aesar) (Important! *see* **Note 5**).
10. Deprotection solution B: 65 % (v/v) TFA, 3 % (v/v) TIPS or TIBS, 2 % (v/v) distilled water, 1 % (w/v) thioanisole, 29 % (v/v) DCM (Important! *see* **Note 5**).
11. Ammonia gas (Air Liquide).
12. Erie Epoxysilane Coated Microarray Slides (Thermo Scientific).
13. Peptide microarray printing solution: Dimethyl sulfoxide (DMSO), phosphate-buffered saline (PBS/0.9 % NaCl in 10 mM phosphate buffer, pH 7.4).

2.2 Kinase Substrate Specificity Screening on Peptide Microarrays

1. Microarray blocking buffer: 1 % bovine serum albumin (BSA) in 100 mM HEPES, pH 7.5.
2. Protein kinase buffer: 5 mM MOPS pH 7.2, 2.5 mM β-glycerolphosphate, 5 mM $MgCl_2$, 1 mM EGTA, 0.4 mM EDTA, 0.5 mM dithiothreitol (DTT) (*see* **Note 6**).
3. Tris-buffered saline (TBS): 50 mM Tris–HCl, 150 mM NaCl, pH 7.5.
4. Tris-buffered saline with Tween (TBST): 0.05 % Tween-20 in TBS.
5. ATP stock solution: 10 mM ATP in distilled H_2O.
6. 0.5 % Sodium dodecyl sulfate (SDS) in distilled H_2O.
7. Pro-Q Diamond phosphoprotein/phosphopeptide microarray stain and destain solution (Life Technologies) (*see* **Note 7**).

2.3 SPOT Synthesis of Peptide Macroarrays and Optimization of Kinase Activity of Selected Peptides

1. Preparation of amino-alkyl ether-modified membranes: 70 % perchloric acid (Alfa Aesar), epibromohydrine (Fluka), 1,3-diaminopropane (Alfa Aesar), 4,7,10-trioxa-1,13-tridecanediamine (Fluka), and sodium methylate (sodium methoxide; Fluka) (*see* **Note 1**).
2. Membrane washing solution: A mixture of 50 ml of MeOH and 1 ml of 70 % aqueous perchloric acid.
3. Membrane activation solution: A mixture of 5 ml epibromohydrine and 500 μl of 70 % aqueous perchloric acid in 1,4-dioxane.
4. Coupling reagents: DIC and N-hydroxybenzotriazole (HOBt; EMD Millipore). Coupling reagents are only necessary when no preactivated amino acid derivatives are used (*see* **Note 4**).
5. Non-preactivated amino acids with protection groups according to the Fmoc strategy [27, 28] (EMD Millipore and GL Biochem); preactivated amino acid derivatives with protection groups according to the Fmoc strategy, e.g., OPfp esters [26] (*see* **Note 4**).
6. Macroarray buffer I: 5 mM MOPS, 2.5 mM β-glycerol-phosphate, 1 mM EGTA, 0.4 mM EDTA, 0.25 mM DTT (*see* **Note 8**).
7. Macroarray buffer II: Macroarray buffer I with 100 mM NaCl, 0.2 mg/ml BSA.
8. Macroarray buffer III: Macroarray buffer I with 100 mM NaCl, 1 mg/ml BSA, 5 mM $MgCl_2$, 50 μM ATP.
9. Macroarray blocking buffer: 5 % sucrose, 4 % skim milk in TBST.
10. Detection antibodies: Generic phosphoserine, phosphothreonine, or phosphotyrosine antibodies.
11. Horseradish peroxidase (HRP)-conjugated secondary antibody recognizing IgGs from the host species that the detection antibody was from.
12. Staining solution I (8.3 ml for 10 ml staining mixture): 100 mg NaCl, 0.5 ml 1 M Tris–HCl, pH 7.4, and 7.8 ml distilled H_2O.
13. Staining solution II (1.7 ml for 10 ml staining mixture): 5 mg 4-chloro-1-naphthol in 1.7 ml methanol. This solution must be prepared fresh shortly before use.
14. Hydrogen peroxide (H_2O_2), 30 %.

3 Methods

3.1 SPOT Synthesis of Free Peptide Amides on Amino-Acid Ester-Modified Cellulose Membranes and Preparation of Peptide Microarrays

The synthesis of the peptides for the preparation of peptide microarrays and macroarrays is carried out by using the SPOT technique [29]. The SPOT technique was initially developed for synthesis of relatively large number of peptides in parallel on the membrane, and the resulting peptide macroarray can then be screened directly [16, 30]. More recently, the SPOT synthesis has been used for the synthesis of large number of free peptides in small amounts [31]. These free peptides are then dissolved and then spotted onto microarray substrates for the preparation of peptide microarrays.

A higher reactivity of the cellulose surface can be achieved by transforming the hydroxyl groups into amino groups, e.g., by esterification of with amino acids [30, 32]. Due to the relative weak ester bond between the cellulose and amino acids, this membrane type is particularly useful for the synthesis of free peptides. The synthesis of the peptides on the cellulose can be carried out by using preactivated amino acids or by performing an in situ activation of the amino acids during the synthesis [19]. The cleavage of the ester bond is possible in strong basic environment (e.g., ammonia gas, hydroxide solutions). The treatment with ammonia gas is preferable, since it results in the free, dry peptides absorbed on the cellulose fibers. The SPOT synthesis can be carried out either automatically or manually. We use the peptide/SPOT synthesizer MultiPep (Intavis Bioanalytical Instruments) for SPOT synthesis. We provide here a description of manual synthesis for a better understanding of the general procedures.

For the synthesis of free peptides, it is recommended to use large spots by applying a reagent volume of 1 μl onto the membrane. To release the peptides from the membrane, punch out the spot areas before or after the final cleavage and transfer the discs into small vials. The peptides can be eluted from the membrane with the microarray print buffer if ammonia gas is used for cleavage. The peptides should already be in solution if basic solutions are used for cleavage.

If not noted differently elsewhere, washing and treatment steps are performed on a rocking shaker and all manual washing steps should be carried out for at least 3 min each time.

3.1.1 Preparation of Esterified Membranes for Free Peptide Synthesis

This procedure is always carried out manually as follows:

1. For the amino functionalization, cut a piece of filter paper to the desired size. For the modification of a sheet with a size of 10 cm × 15 cm (that fits on a tray of the synthesizer and can accommodate 9 × 12 = 118 spots), prepare the 12.5 ml of amino functionalization solution. For larger membranes, use more reagent solutions accordingly.

2. Transfer the mixture into a chemically resistant box with lid (stainless steel or polypropylene) and place the filter paper into the liquid. Avoid air bubbles trapped underneath the paper. The membrane should be completely soaked in the solution.
3. Let the membrane react with the reaction mixture in the closed box overnight (*see* **Note 9**).
4. After the treatment, wash the membrane at least three times with DMF. If necessary the membrane can be stored at −20 °C for a few weeks until needed (*see* **Note 10**).
5. For the Fmoc deprotection, treat the membrane twice with piperidine solution for at least 5 min each time.
6. Wash the membrane at least four times with DMF, followed by washing at least twice with MeOH or EtOH.
7. To perform an optional staining [32] (*see* **Note 11**): Treat the membrane with staining solution for at least 2 min until the filter paper shows a homogeneous blue color. If the staining is insufficient, repeat with fresh staining solution.
8. After staining, wash the membrane at least twice with MeOH or EtOH, until the liquid remains colorless.
9. After thorough air-drying, the membrane is ready for the first coupling (*see* **Note 12**).

3.1.2 Preparation of Coupling Solutions

Generally, two different methods can be used for the preparation of coupling solutions.

The first method is using preactivated Fmoc-protected amino acid derivatives (e.g., pentafluorophenyl esters) [16]. The advantage of this method is the use of a single type of reagent, which simplifies the preparation of the solutions and reduces the chances of human errors. Unfortunately, activated esters are commercially available for standard amino acids only. Preactivated derivatives of other amino acids have to be synthesized in advance [26]. This second approach is in situ activation of protected Fmoc-amino acid derivatives. Since for the synthesis of free peptides for the microarray production only standard amino acids are used, we only describe the first method in this chapter.

Most solutions of preactivated amino acids are prepared by dissolving the corresponding amino acid derivatives in NMP at a concentration of 0.3 M. Due to poor solubility, the serine derivative must be dissolved in amine-free DMF. Except for the arginine derivative, all solutions can be used for at least 1 week if the stock solutions are stored at or below −20 °C. Replace amino acid solutions of the previous day with fresh amounts from the stock solutions each day (*see* **Note 4**). Due to the instability of the dissolved preactivated arginine derivative, it must be prepared fresh every day.

3.1.3 SPOT Synthesis on the Cellulose Membrane

This protocol describes the SPOT synthesis on large spots. Usually, a pipetting robot is used to deliver the coupling solutions; but for a small number of large spots the solution can be pipetted manually.

To locate the large spots after synthesis, mark the center of each spot at the start or during the synthesis with a pencil. For peptide macroarrays, the edges of the entire area should be marked in the same way. Since the application of 1 μl of coupling solution results in spots with a diameter of about 7 mm, the distance between the centers of two spots should be at least 8 mm.

1. For amino acid coupling, the synthesis runs from the C- to the N-terminus. Deliver the desired volumes of activated amino acid solutions to the corresponding positions on the membrane. For the first coupling cycle, use 1 μl amino acid solution (large spots for microarray production) or about 0.1 μl (small spots for macroarrays). For all other steps, use 20 % more of amino acid solution in order to cover the entire spot area.
2. After the delivery cycle allow the reagents to react for at least 20 min.
3. A repeat of the spotting is recommended in order to achieve a higher coupling yield.
4. Capping: This step reduces the number of side products by acetylation of unreacted amino groups. At the first coupling cycle place the membrane face down in a box filled with an appropriate amount of capping solution. Do not leave air bubbles under the membrane. Do not shake!
5. Renew the capping solution after about 5 min and let it react for another 20 min.
6. For all subsequent cycles treat the membrane with capping solution twice for 5 min.
7. To remove the Fmoc-protecting group, wash the membrane with DMF four times.
8. Treat with piperidine solution twice for 5 min each time.
9. Wash at least four times with DMF.
10. Wash at least twice with MeOH or EtOH. Do not perform this step for the final Fmoc removal.
11. Staining (optional) [33] (*see* **Note 11**): Stain the membrane with staining solution in a box while shaking and leave it in until the spots are stained sufficiently. If necessary, renew the solution (*see* **Note 13**).
12. For destaining, wash the membrane at least twice with MeOH or EtOH until the wash solution remains colorless.
13. Air-dry the membrane before continuing with the next coupling cycle (*see* **Note 12**).

14. Building up the peptide chain: Except for the last coupling cycle, repeat **steps 1–12**. For the last coupling cycle, carry out only **steps 1** and 7!
15. Upon the removal of the last Fmoc-protecting groups, wash the membrane at least four times with DMF followed by washes with DCM three times.
16. Air-dry the membrane.
17. Final side-chain deprotection: Treat the membrane with 20 ml of deprotection solution A. The membrane must always be submerged in the deprotection solutions. Keep the box tightly closed. Do not shake!
18. After 30 min, pour out the solution very carefully (*see* **Note 14**). If the membrane is already soft, remove the residual solution completely with a pipette.
19. Treat the membrane with at least 20 ml of deprotection solution B for 3 h in the closed box without shaking.
20. Remove the solution carefully with a pipette.
21. Wash the membrane first with DCM at least five times (*see* **Note 14**) and DMF twice followed by MeOH twice.
22. Dry the membrane in a fume hood.

3.1.4 Cleavage of the Peptides from Membrane as Free Peptide Amides

The method described here involves the exposure of the entire dry membrane or the punched-out spots to ammonia gas that breaks the ester bond between the peptides and cellulose and forms a C-terminal amide (*see* **Note 15**).

1. Place the dry membrane or the punched-out spots in a glass desiccator (*see* **Note 16**).
2. Set the desiccator under vacuum.
3. Fill the desiccator with ammonia gas.
4. To replace most of the air by ammonia repeat **steps 2** and **3** at least twice.
5. Let the reaction proceed overnight.
6. Open the desiccator under a fume hood (Attention! ammonia gas is highly corrosive and irritant!). Let the gas dissipate for at least 30 min.
7. Transfer the discs into wells of microtiter plates (MTPs) or vials followed by addition of adequate amount of DMSO to dissolve the adsorbed free peptides (*see* **Note 17**). The resulting peptide stock solutions can be kept at −20 °C for long-term storage.

The amounts of peptides on amino acid ester-modified membranes for preparing peptide microarrays should range from 340 to 400 nmol/cm^2. If a common hole puncher is used, the peptide amount recovered typically ranges from 100 to 120 nmol. The purity of the peptides may vary significantly, ranging from 40 to 91 % [34, 35], depending on their sequences [36] (*see* Fig. 1).

3.1.5 Preparation/ Printing of Peptide Microarrays

To prepare peptides for microarray printing, the peptide stock solutions are diluted with three times PBS and 15 μl of each is plated into 384-well microtiter plates according to the final layout on the arrays. The epoxysilane-coated microarray slides from Thermo-Fisher are employed as the substrates for spotting without pretreatments. Printing is carried out by a microarrayer equipped with quill pins for peptide delivery. The printed arrays are left on the arrayer overnight under the relative humidity of 50 %. Orientation markers and appropriate control peptides are integrated for quality assessment and array orientation. Printed peptide microarrays are stored in vacuum at −80 °C until use.

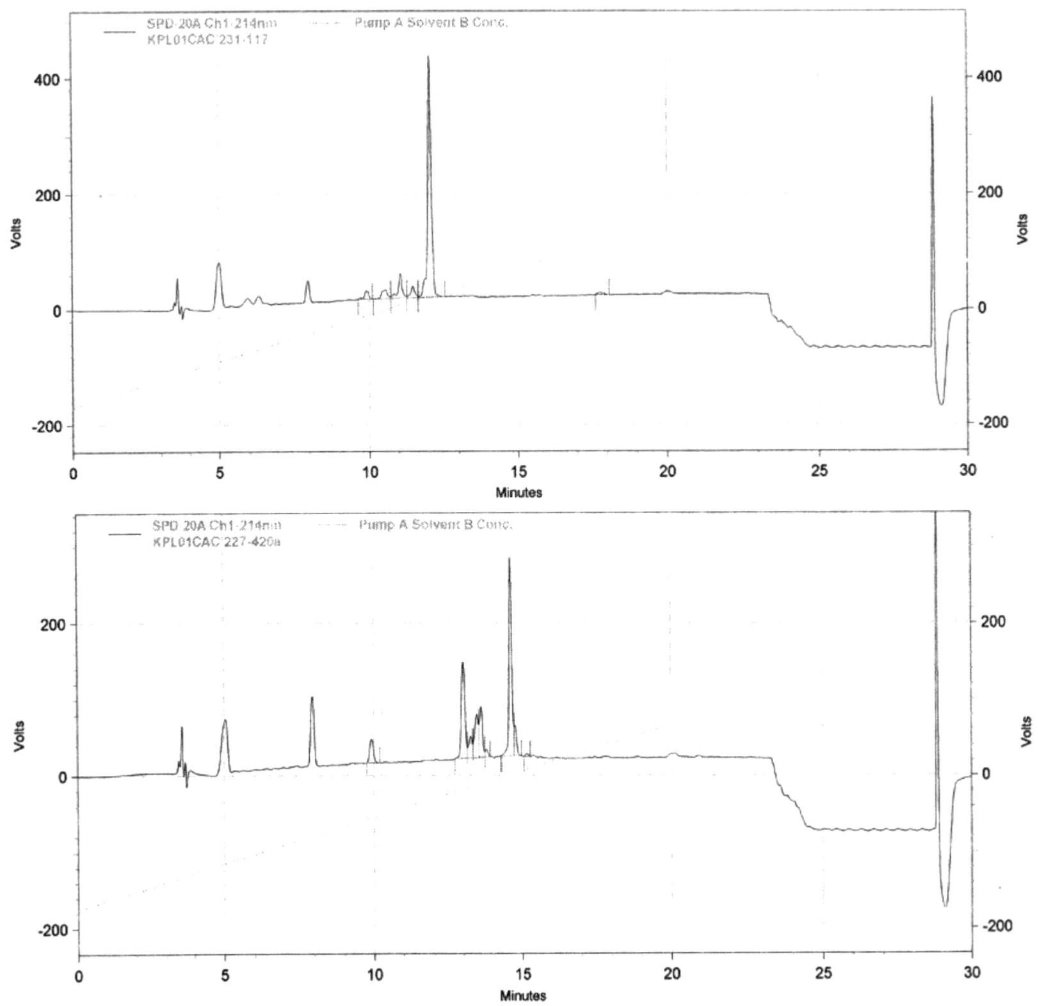

Fig. 1 Example of two HPLC images representing different kinase substrate peptides: The top image shows an optimum purity, while the bottom one represents a more typical example of purity

3.2 Kinase Substrate Specificity Screening on Peptide Microarrays

In the following protocol, we have generalized the conditions to make it suitable for most purified protein kinases that have detectable activity in vitro. The concentration of the kinase, reaction buffer, and the incubation time for the kinase assay on a peptide microarray slide can vary from kinase to kinase and may depend on the specific activity of the kinase. We recommend a negative control experiment performed in parallel without the kinase or with a kinase-dead mutant. A phosphoprotein/phosphopeptide sensor dye, Pro-Q Diamond stain can be used for probing with high sensitivity and low false-positive rate [37]. The consensus substrate sequence identified from microarray data analysis will be used for further optimization or individual verification.

1. Block the slide in a microarray reaction tray with the microarray blocking buffer for 1 h at room temperature with agitation (*see* **Note 18**).
2. Wash the slide with TBST three times, 2 min each time.
3. Take out the slide from TBST using forceps, and allow excess liquid to slip off. Clip on to the slide a microarray incubation chamber.
4. Make up a proper volume of kinase reaction mix by diluting purified active protein kinase in protein kinase buffer, and supplement with 100 μM ATP. Load the solution into the incubation chamber.
5. Incubate the reactions for 2 h at 30 °C in a shaking incubator. Keep the microarray in a humidity chamber if necessary (*see* **Note 19**).
6. Carefully remove the incubation chamber. Wash the slide with 0.5 % SDS twice, 5 min each time.
7. Wash the slide with TBST seven times, 5 min each time to remove SDS.
8. Block the slide in the microarray blocking buffer again for 1 h at room temperature.
9. Stain the slide with Pro-Q Diamond stain for 1 h at room temperature in the dark.
10. Destain with the Pro-Q Diamond destain solution three times, 15 min each time.
11. Rinse the slide in H_2O once. Wash in H_2O for 15 min.
12. Dry the slide using N_2 flow (*see* **Note 20**).
13. Scan in a microarray scanner at a wavelength of 532 nm or 543 nm (*see* **Note 21**). Examples of obtained images are presented in Fig. 2.
14. Use microarray image analysis software to archive quantitative data and generate a consensus peptide sequence.

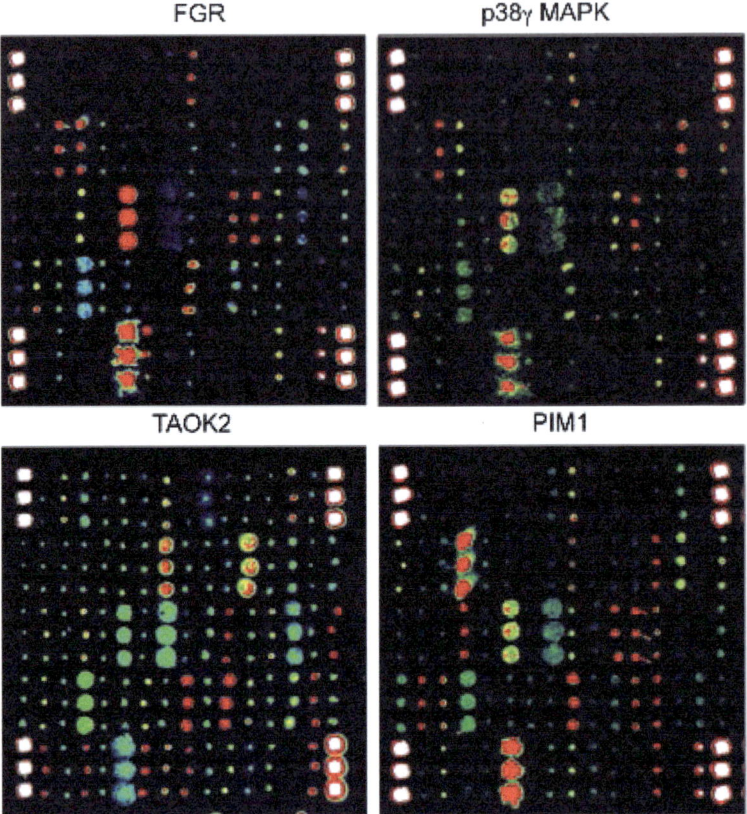

Fig. 2 Examples of protein kinase substrate peptide microarray images: Each kinase substrate peptide was printed in triplicate and phosphorylated by preparations of purified active protein kinases on epoxysilane-coated microarray slides. The phosphorylation of the peptides was visualized with Pro-Q Diamond stain. In these images, the first grid of peptide spots on peptide microarrays phosphorylated in vitro by the protein kinases FGR,p38-gamma MAP kinase,TAOK2, and Pim1 are shown for comparison. Orientation markers are printed in each of the four corners

3.3 SPOT Synthesis of Peptide Macroarrays for Substrate Sequence Optimization

Peptide macroarrays are a very useful tool for the optimization of lead sequences identified from peptide microarrays or phage libraries [38]. The use of microarrays may be considered if the amount of the kinase is limited. Several strategies are described for the optimization of peptides on peptide macroarrays, e.g., substitution analysis (systematic substitution of all residues of the sequence by other amino acids), truncation analysis (systematic reduction of the length of a peptide), and loop scan (systematic variation/insertion of cyclization of a peptide) [20]. Signals can be detected by radioactivity [39, 40], fluorescence [40], chemiluminescence, or colorimetric staining. In this chapter, we only describe the colorimetric detection due to the simplicity in setup and the lower background compared to fluorescence-based detection as result of autofluorescence of peptides.

3.3.1 Preparation of Amino-Alkyl Ether-Linked Membranes for Probing Directly on the Membrane as Peptide Macroarrays

In contrast to those modified by amino acids linked via an ester bond, for the SPOT synthesis of peptide macroarrays the use of amino-alkyl ether-linked membranes [19, 20] is recommended, since they keep the peptides stably attached to the cellulose. This stability also allows the storage at −20 °C for several months. For that reason, it makes sense to prepare a large membrane (e.g., 19 cm × 29 cm) and cut it into several smaller pieces if needed.

1. Wash the membrane with about 50 ml of membrane washing solution for several minutes. Then air-dry the membrane.
2. Activate the membrane by treatment with 50 ml of membrane activation solution in a closed box.
3. After 3 h, wash the membrane once with EtOH or MeOH for about 15 min.
4. Leave the activated membrane above in about 60 ml of a 50–75 % solution of 1,3-diaminopropane (CAPE) or 4,7,10-trioxa-1,13-tridecanediamine (TOTD, trioxa, DIPEG) in DMF overnight. The concentration of the amines affects the density of the amino groups on the membrane surface.
5. On the next day, wash the membrane three times with DMF, twice with water, and then three times with MeOH [41, 42].
6. After treatment with a methanolic suspension of 5 M sodium methylate, wash the membrane with MeOH three times and water five times followed by MeOH three times.
7. Staining (optional) [33]: Wash the membrane at least three times with MeOH or EtOH for at least 30 s each.
8. Treat the membrane with staining solution for at least 2 min until the filter paper shows a homogeneous blue color (*see* **Note 22**). If staining is insufficient, renew the staining solution.
9. After staining, wash the membrane at least twice with MeOH or EtOH, until the wash solution remains colorless.
10. Dry the membrane with the airstream of a fume hood (*see* **Note 12**). The membrane is ready for the first coupling or storage.
11. For the direct use, cut the required size of the membrane and store the remaining sheet. The amino functionality of amino-alkyl ether-linked membranes is typically in the range of 350–700 nmol/cm^2.

3.3.2 Preparation of Coupling Solutions

As mentioned in Subheading 3.1.2, two different methods are commonly used for the preparation of coupling solutions. The first method uses Fmoc-protected amino acid pentafluorophenyl esters, which has been described in detail elsewhere [15]. For nonstandard amino acids, e.g., phosphoamino acid derivatives, the preactivated derivatives need to be synthesized. Alternatively, an in situ activation of protected Fmoc-amino acid derivative can be used. We elaborate further on the latter method here, as the inclusion of phosphopeptides as positive controls is usually necessary in kinase activity assays.

1. Dissolve Fmoc-amino acids in 0.9 M HOBt solution (in NMP) to a concentration of 0.45 M. Except for the arginine derivatives, these solutions can be stored at −20 °C for at least a week. Use fresh aliquots of the prepared amino acid/HOBt solutions on a daily basis (*see* **Note 4**).
2. To these solutions add a fresh prepared mixture of 20 % DIC in NMP at a ratio of 3:1 (e.g., for final volume of 100 μl of in situ-activated amino acid solution mix 75 μl of HOBt/amino-acid solution with 25 μl of 20 % DIC/NMP) (*see* **Note 23**). This solution is ready to use.

3.3.3 SPOT Synthesis of the Peptide Macroarray

The SPOT synthesis of a peptide macroarray follows a similar protocol as described in Subheading 3.1.3. Due to the much smaller volume delivered to each spot (0.1 μl instead of 1.0 μl) and the large number of peptides, manual synthesis of peptide macroarrays is not recommended. At minimum, delivery of the amino acid solutions should be carried out by a robotic system. Since the coupling protocol may vary between the different spotting devices, no general protocol is described here but the instructions are available in Subheading 3.3.2 above. Ether-modified membranes are usually a bit sturdier than esterified membranes, which makes them much easier to handle during the final TFA treatment (Subheading 3.1.3, **step 17**). After this step, the membrane is ready for use or can be stored at or below −20 °C.

3.3.4 Probing of the Peptide Macroarray

If the synthesized macroarray is to be used immediately after the TFA treatment, skip the drying step. Otherwise, soak the dried membrane in MeOH or EtOH for 10 min prior to probing.

1. Block the membrane in macroarray buffer II at room temperature overnight.
2. Incubate the membrane in macroarray buffer III for 1 h at 30 °C in a shaking incubator.
3. Dilute the purified active protein kinase in the kinase assay buffer supplemented with 100 μM ATP. The final concentration of the kinase varies according to its specific activity.
4. Incubate the membrane in the solution for 0.5–2 h at 30 °C in a shaking incubator.
5. Wash the membrane in TBST three times, 10 min each time at room temperature.

6. Block unspecific binding sites with the macroarray blocking buffer for 2 h at room temperature.
7. Wash the membrane once in the wash buffer for 5 min.
8. Make up selected generic phospho-Ser/Thr/Tyr antibody with macroarray blocking buffer following the supplier's recommendation.
9. Incubate the membrane in the antibody solution overnight at 4 °C with agitation.
10. Wash the membrane in wash buffer three times, 10 min each time.
11. Incubate the membrane with the HRP-conjugated secondary antibody in macroarray blocking buffer for 2 h at room temperature.
12. Wash the membrane in wash buffer three times, 10 min each time.
13. Wash the membrane in TBS three times, 5 min each time.
14. Mix staining solution I with staining solution II. Immediately before use, add 5 μl of 30 % H_2O_2.
15. Treat the membrane with the above staining mixture until the signals are fully developed.
16. Stop the reaction by washing thoroughly with water. Scan the wet membrane using a flatbed scanner and save the resulting image (*see* **Note 24**). An example of macroarray image generated is shown in Fig. 3.

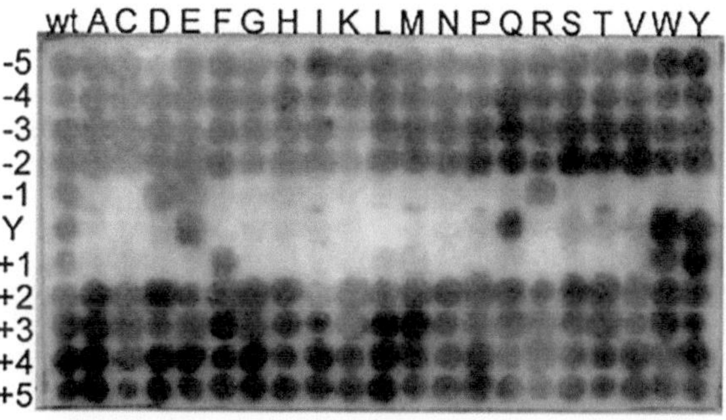

Fig. 3 Example of a substitutional analysis on a protein kinase peptide substrate macroarray: From a wild-type substrate peptide sequence of human JAK kinase (first column on the macroarray), amino acids at each position were substituted with other amino acids systematically. Active JAK kinase was used to phosphorylate the macroarray membrane. The phosphorylation signal of the each peptide spot was detected with the 4G10 monoclonal phosphotyrosine-specific antibody as described in Subheading 3.3

4 Notes

1. Solvents and reagents used here are toxic, corrosive, irritant, and/or flammable and can be hazardous. Therefore, precautions while handling them should be taken.
2. The quality of solvents for washing steps should be at least of ACS grade. Solvents for dissolving reagents for the synthesis must be amine and water free. Due to the risk of decomposition under the influence of light, organic solvents, with the exception of MeOH and EtOH, should be stored in the dark. The water used should always be distilled/deionized.
3. Due to linear structure of the molecule and the resulting flexibility, β-alanine is commonly used for amine functionalization of the filter paper by esterification. More recently, due to the higher loading on the membrane and lower stability during cleavage, glycine is used to functionalize membranes for free peptide synthesis [32]. While other amino acids may be used, a higher risk of losing functionality is expected as a result of the lower chemical stability of the ester bond between the amino acid and cellulose. The amino functionalization solution is prepared prior to use.
4. Reagents must be protected from moisture. To avoid condensation from air humidity, reagent containers should be kept unopened for approximately 30 min after storage in fridge or freezer before use.
5. Do not pour the mixture into the TFA! Otherwise, it could heat up to a dangerous level! Mix the additives first and then pour the TFA into that mixture.
6. Protein kinase buffer is usually made as 5× or 10× stock solution without DTT and stored at −20 °C. DTT is dissolved in DMSO to a stock concentration of 1 M and frozen in small aliquots. Before use, dilute the 5× or 10× protein kinase buffer with dH_2O and add DTT to a final concentration of 0.5 mM.
7. Pro-Q Diamond destaining solution can also be made with 50 mM sodium acetate in 20 % acetonitrile (v/v), pH 4.
8. Macroarray buffer I is usually made as 5× or 10× stock solution without DTT and stored at −20 °C. DTT should be freshly added before use.
9. For more consistent physicochemical behavior, it is recommended after the functionalization overnight to treat the membrane with an appropriate amount of capping solution for at least 20 min.
10. Loss of functionality is possible after long-term storage. For long-term storage, −80 °C is recommended. Wash the membrane at least three times with MeOH (or EtOH) and let it

air-dry prior to storage. Any traces of amines or traces of water on the membrane could lead to complete loss of activity within a relative short period of time! Place the membrane in a sealed plastic bag. To proceed with the synthesis after storage, treat the membrane once with DMF for about 20 min.

11. The staining has no detrimental effect on peptide synthesis process itself. Free amino groups of coupled amino acids are stained blue. Thus, it is an indicator for the completion of the previous coupling step. Since unreacted amino groups are acetylated during the capping step, the spots remain colorless.
12. For quick drying, the membrane can be further washed twice with DEE.
13. The staining of the spots must not be too strong, since a high amount of absorbed BPB could lead to an incorporation of some dye into the peptide. In that case, the removal of the dye is difficult and may affect detection after incubation. Due to varying acidity of the coupled amino acids and the built-up peptide chain, differences in the intensity of the staining of the spots are normal: for example, spots where aspartic acid or cysteine as last coupled amino acids may show no or very little staining, whereas alanine and lysine show usually a deep blue color.
14. During TFA treatment, the membrane may become very soft. Do not pour the liquid directly onto the membranes as it may destroy it. Shake very gently and do not try to lift the membrane out until, after several washing steps, it becomes harder and less likely to be torn.
15. If it is necessary to yield a free carboxyl group at the C-terminus, do not treat the membrane with ammonia gas. Instead, punch out the spots, transfer them into vials, and treat them with an aqueous strong basic solution such as 50 % ammonium hydroxide, 1 % triethylamine, or 1 M sodium hydroxide solutions [18, 39]. This solution needs to be exchanged for microarray printing buffer before microarray production.
16. Since many desiccators made from plastics are not inert to ammonia gas, it is strongly recommended to use glass desiccators only.
17. To reduce possible contamination with side products that are in higher concentration at the rim of the spot, the diameter of the punched-out membrane discs should be smaller than the spot diameter.
18. All incubations and washes are performed in microarray reaction trays on a horizontal shaker or tilt rocking platform at low speed.

19. A humidity chamber keeps the slide from drying up during the 2-hour incubation, especially when the reaction volume is small. We suggest the speed of the shaking incubator be set below 100 rpm. High shaking speed could cause increase of background.
20. If using N_2 flow, dry the slide quickly to avoid water markers in the background when scanning. An alternative method is to dry the slide by spinning in a swing-bucket centrifuge.
21. Pro-Q Diamond stain has excitation maxima of 555 nm and emission maxima of 580 nm [37]. Please refer to the user manual of Pro-Q Diamond phosphoprotein/phosphopeptide microarray stain from Molecular Probes for detailed information.
22. In contrast to esterified membranes, the background of TODT and CAPE membranes from the dye staining may disappear slowly over several coupling cycles. Nevertheless, it should not affect the quality of the synthesis.
23. To avoid clogging of the needle by urea during automatic spotting, shake the in situ-activated amino acid solutions for about 30 min; then centrifuge the mixtures and transfer the supernatants into vials of the synthesis rack.
24. Scan the membrane when it is wet. The staining could fade during drying and change color, making the spots undistinguishable.

References

1. Manning G, Whyte DB, Martinez R et al. (2002) The protein kinase complement of the human genome. Science 298:1912–1934
2. Songyang Z, Cantley LC (1995) Recognition and specificity in protein tyrosine kinase-mediated signalling. Trends Biochem Sci 20:470–475
3. Songyang Z, Lu KP, Kwon YT et al. (1996) A structural basis for substrate specificities of protein Ser/Thr kinases: primary sequence preference of casein kinases I and II, NIMA, phosphorylase kinase, calmodulin-dependent kinase II, CDK5, and Erk1. Mol Cell Biol 16:6486–6493
4. Faux MC, Scott JD (1996) More on target with protein phosphorylation: conferring specificity by location. Trends Biochem Sci 21:312–315
5. Kreegipuu A, Blom N, Brunak S (1998) Statistical analysis of protein kinase specificity determinants. FEBS Lett 430:45–50
6. Songyang Z (1999) Recognition and regulation of primary-sequence motifs by signaling modular domains. Prog Biophys Mol Biol 71:359–372
7. Kemp BE, Benjamini E, Krebs EG (1976) Synthetic hexapeptide substrates and inhibitors of 3′:5′-cyclic AMP-dependent protein kinase. Proc Natl Acad Sci U S A 73:1038–1042
8. Kemp BE, Graves DJ, Benjamini E et al. (1977) Role of multiple basic residues in determining the substrate specificity of cyclic AMP-dependent protein kinase. J Biol Chem 252:4888–4894
9. Casnellie JE, Krebs EG (1984) The use of synthetic peptides for defining the specificity of tyrosine protein kinases. Adv Enzyme Regul 22:501–515
10. Safaei J, Manuch J, Gupta A, Stacho L, Pelech S (2011) Prediction of 492 human protein kinase substrate specificities. Proteome Sci 9(Suppl 1):S6
11. Tegge W, Frank R, Hofmann F et al. (1995) Determination of cyclic nucleotide-dependent protein kinase substrate specificity by the use of peptide libraries on cellulose paper. Biochemistry 34:10569–10577
12. MacBeath G, Schreiber SL (2000) Printing proteins as microarrays for high-throughput function determination. Science 289:1760–1763

13. Houseman BT, Huh JH, Kron SJ et al. (2002) Peptide chips for the quantitative evaluation of protein kinase activity. Nat Biotechnol 20: 270–274
14. Reimer U, Reineke U, Schneider-Mergener J (2002) Peptide arrays: from macro to micro. Curr Opin Biotechnol 13:315–320
15. Frank R (2002) The SPOT-synthesis technique. Synthetic peptide arrays on membrane supports—principles and applications. J Immunol Methods 267:13–26
16. Kramer A, Schneider-Mergener J (1998) Synthesis and application of peptide libraries bound to continuous cellulose membranes. Methods Mol Biol 87:25–39
17. Reineke U, Volkmer-Engert R, Schneider-Mergener J (2001) Applications of peptide arrays prepared by the SPOT-technology. Curr Opin Biotechnol 12:59–64
18. Blackwell HE (2006) Hitting the SPOT: small-molecule macroarrays advance combinatorial synthesis. Curr Opin Chem Biol 10:203–212
19. Hilpert K, Winkler DFH, Hancock REW (2007) Peptide arrays on cellulose support: SPOT synthesis—a time and cost efficient method for synthesis of large numbers of peptides in a parallel and addressable fashion. Nat Protoc 2:1333–1349
20. Winkler DFH, Campbell WD (2008) The Spot technique: Synthesis and screening of peptide macroarrays on cellulose membranes. Methods Mol Biol 494:47–70
21. Schutkowski M, Reineke U, Reimer U (2005) Peptide arrays for kinase profiling. Chembiochem 6:513–521
22. Diks SH, Kok K, O'Toole T et al. (2004) Kinome profiling for studying lipopolysaccharide signal transduction in human peripheral blood mononuclear cells. J Biol Chem 279:49206–49213
23. Schutkowski M, Reimer U, Panse S et al. (2004) High-content peptide microarrays for deciphering kinase specificity and biology. Angew Chem Int Ed Engl 43:2671–2674
24. Uttamchandani M, Chan EW, Chen GY et al. (2003) Combinatorial peptide microarrays for the rapid determination of kinase specificity. Bioorg Med Chem Lett 13:2997–3000
25. Wildemann D, Erdmann F, Alvarez BH et al. (2006) Nanomolar inhibitors of the peptidyl prolyl cis/trans isomerase Pin1 from combinatorial peptide libraries. J Med Chem 49:2147–2150
26. Atherton E, Sheppard RC (1989) 7.2. Activated esters of Fmoc-amino acids. In: Solid phase peptide synthesis—a practical approach. IRL press at Oxford University Press, Oxford, UK, pp 76–78
27. Fields GB, Noble RL (1990) Solid phase synthesis utilizing 9-fluorenylmethoxycarbonyl amino acids. Int J Pept Protein Res 35: 161–214
28. Zander N, Gausepohl H (2002) Chemistry of Fmoc peptide synthesis on membranes. In: Koch J, Mahler M (eds) Peptide Arrays on Membrane Support. Springer, Berlin, Heidelberg, pp 23–39
29. Hilpert K, Winkler DFH, Hancock REW (2007) Cellulose-bound peptide arrays: Preparation and applications. Biotechnol Genet Eng Rev 24:31–106
30. Frank R (1992) Spot-synthesis: An easy technique for the positionally addressable, parallel chemical synthesis on a membrane support. Tetrahedron 48:9217–9232
31. Volkmer R, Kretzschmar I, Tapia V (2012) Mapping receptor–ligand interactions with synthetic peptide arrays: Exploring the structure and function of membrane receptors. Eur J Cell Biol 91:349–356
32. Kamradt T, Volkmer-Engert R (2004) Cross-reactivity of T lymphocytes in infection and autoimmunity. Mol Divers 8:271–280
33. Krchnak V, Wehland J, Plessmann U et al. (1988) Noninvasive continuous monitoring of solid phase peptide synthesis by acid-base indicator. Collect Czech Chem Comm 53: 2542–2548
34. Molina F, Laune D, Gougat C et al. (1996) Improved performances of spot multiple peptide synthesis. Pept Res 9:151–155
35. Portwich M, Keller S, Strauss HM et al. (2007) A network of coiled-coil associations derived from synthetic GCN4 leucine-zipper arrays. Angew Chem Int Ed 46:1654–1657, Supporting Information
36. Kramer A, Reineke U, Dong L et al. (1999) Spot-synthesis: observations and optimizations. J Pept Res 54:319–327
37. Martin K, Steinberg TH, Cooley LA et al. (2003) Quantitative analysis of protein phosphorylation status and protein kinase activity on microarrays using a novel fluorescent phosphorylation sensor dye. Proteomics 3:1244–1255
38. Petter C, Scholz C, Wessner H et al. (2008) Phage display screening for peptidic chitinase inhibitors. J Mol Recognit 21:401–409
39. Lizcano JM, Deak M, Morrice N et al. (2002) Molecular basis for the substrate specificity of NIMA-related kinase-6 (NEK6). Evidence that Nek6 does not phosphorylate the hydrophobic motif of ribosomal S6 protein kinase and serum-

and glucocorticoid-induced protein kinase in vivo. J Biol Chem 277:27839–27849

40. Panse S, Dong L, Burian A et al. (2004) Profiling of generic anti-phosphopeptide antibodies and kinases with peptide microarrays using radioactive and fluorescence-based assays. Mol Divers 8:291–299

41. Licha K, Bhargava S, Rheinlander C et al. (2000) Highly parallel nano-synthesis of cleavable peptide-dye conjugates on cellulose membranes. Tetrahedron Lett 41:1711–1715

42. Ast T, Heine N, Germeroth L et al. (1999) Efficient assembly of peptomers on continuous surfaces. Tetrahedron Lett 40:4317–4318

Chapter 15

Rapid Identification of Protein Kinase Phosphorylation Site Motifs Using Combinatorial Peptide Libraries

Chad J. Miller and Benjamin E. Turk

Abstract

Eukaryotic protein kinases phosphorylate substrates at serine, threonine, and tyrosine residues that fall within the context of short sequence motifs. Knowing the phosphorylation site motif for a protein kinase facilitates designing substrates for kinase assays and mapping phosphorylation sites in protein substrates. Here, we describe an arrayed peptide library protocol for rapidly determining kinase phosphorylation consensus sequences. This method uses a set of peptide mixtures in which each of the 20 amino acid residues is systematically substituted at nine positions surrounding a central site of phosphorylation. Peptide mixtures are arrayed in multiwell plates and analyzed by radiolabel assay with the kinase of interest. The preferred sequence is determined from the relative rate of phosphorylation of each peptide in the array. Consensus peptides based on these sequences typically serve as efficient and specific kinase substrates for high-throughput screening or incorporation into biosensors.

Key words Protein kinases, Peptide libraries, Enzyme specificity, Substrate profiling, Kinase assay, Signal transduction

1 Introduction

Protein kinases target specific downstream substrates in cells through a combination of direct and indirect physical interactions [1]. Kinases interact directly with their substrates through both "proximal" interactions between the kinase active site and the site of phosphorylation, as well as more "distal" non-catalytic interactions to substrate docking sites. Most kinases phosphorylate substrates in the context of specific sequence motifs that interact favorably with the kinase catalytic cleft [2]. Early studies with peptides derived from the phosphorylation sites of known protein substrates established the idea of the consensus sequence, in which key residues proximal to the phosphoacceptor residue were particularly critical for efficient phosphorylation [3]. These consensus sequences were shown to vary among different kinases (Table 1). More recently, the application of peptide library methods to

Hicham Zegzouti and Said A. Goueli (eds.), *Kinase Screening and Profiling: Methods and Protocols*, Methods in Molecular Biology, vol. 1360, DOI 10.1007/978-1-4939-3073-9_15, © Springer Science+Business Media New York 2016

Table 1
Consensus sequences recognized by protein kinases

Kinase	Consensus sequence(s)	References
cAMP-dependent protein kinase (PKA)	R-R/K-x-S-ϕ	[25]
AKT	R-x-R-x-x-S/T-ϕ	[26]
AMP-activated protein kinase (AMPK)	ϕ-x-R-x-x-S-x-x-x-I/L	[27, 28]
Mitogen-activated protein kinases (MAPKs)	P/ϕ-x-S/T-P	[29]
Cyclin-dependent kinases (CDKs)	S/T-P-x-K/R	[5]
Casein kinase 1 (CK1)	D/E-D/E-D/E-x-x-S/T-ϕ pS/pT-x-x-S/T-ϕ	[30]
Casein kinase 2 (CK2)	S/T-D/E-x-D/E	[6]
NimA-related kinase (NEK)	ϕ-x-x-S/T	[31]

x: any amino acid
Underline: phosphorylated residue
ϕ: hydrophobic residue
pS/pT: priming phosphoserine/phosphothreonine

analyzing kinase specificity, as well as mapping of large numbers of phosphorylation sites on protein substrates, has suggested that the simple notion of the consensus sequence is inadequate to define the substrate specificity of a kinase [4–7]. For example, a kinase may prefer, though not absolutely require, certain residues or sets of residues at a particular position relative to the phosphorylation site. In addition, at some positions a kinase may be permissive for most residues, and yet strongly disfavor others [8, 9]. From this standpoint the sequence specificity of a kinase may be better described quantitatively by a weight matrix (also called a position-specific scoring matrix) that provides the relative rates of phosphorylation of each residue at multiple positions near the site of phosphorylation [4]. Such matrices are often depicted graphically as heat maps or sequence logos [10, 11].

Understanding the phosphorylation site specificity of a kinase has several applications in biochemistry, biology, and drug discovery. Such information facilitates the design of peptides that are typically highly efficient substrates for the kinase [12, 13]. Optimized peptides can be used as substrates in high-throughput screening to identify kinase inhibitors, and can substantially reduce the quantity of kinase required for screening in comparison to generic peptide or protein substrates, which tend to be phosphorylated with low efficiency. In addition, optimized substrate sequences can be incorporated into "biosensors" that can be used to read out the activity of kinases in living cells in spatial and temporal dimensions [14]. Lastly, knowing sequences that are preferred by a kinase facilitates identification of new phosphorylation sites by scanning sequences of known protein substrates or by searching protein sequence or mass spectrometry databases to identify novel substrates. Several bioinformatics tools have been developed to identify kinase substrates by searching such databases using peptide library-derived weight matrices of substrate specificity [4, 15–17].

Multiple methods have been developed to determine sequence motifs phosphorylated by kinases. Early peptide library approaches used either mixture-based "oriented" peptide libraries that were analyzed by Edman sequencing or peptides immobilized in arrays on cellulose membranes [5, 18, 19]. Recent developments have provided advantages in more reliably identifying negatively selected residues and decreasing background associated with immobilized substrates. Current methods include analysis of proteome-derived peptide libraries by mass spectrometry, and tagging peptides bound to beads by thiophosphorylation with ATP-γ-S [20–22]. Here, we describe a positional scanning peptide library (PSPL) method developed in our laboratory [23, 24]. This method employs one of two "universal" sets of 198 biotinylated peptide mixtures that can be used to generally profile either serine-threonine or tyrosine kinases. Peptides are phosphorylated in solution with radiolabeled ATP, and then captured on streptavidin-coated membranes. Following washing and drying of the membrane, radiolabel incorporation into each peptide mixture is quantified by phosphor imaging. This method provides the relative phosphorylation rate of peptides incorporating every amino acid at several positions surrounding the phosphoacceptor site. In this way, the method indicates which residues are "essential," which are preferred, and which are disfavored at each position. Quantified data are readily converted into a weight matrix that can be used for protein database scanning.

2 Materials

2.1 Positional Scanning Peptide Library

1. The Positional Scanning Peptide Library (PSPL) for analysis of Ser-Thr kinases is made up of 198 peptide mixtures having the general sequence Y-A-X-X-X-X-X-S/T-X-X-X-X-G-A-K-K (biotin). For each peptide mixture, eight of the X positions represent a degenerate mixture of all amino acids (except Cys, Ser, and Thr), S/T indicates an even mixture of Ser and Thr, and K(biotin) is ε-(N-biotinyl-6-aminohexanoyl)-lysine. The remaining X is fixed as 1 of the 20 standard amino acids, phosphothreonine (pT), or phosphotyrosine (pY). These 22 fixed residues at 9 positions yield the 198 peptide mixtures ($22 \times 9 = 198$) that constitute the PSPL (*see* **Note 1**). This library is commercially available from Anaspec, Inc. (sold as Kinase Substrates Library, Groups I and II).

2. Dimethyl sulfoxide (DMSO).
3. Argon gas.
4. 20 mM HEPES, pH 7.4.
5. Distilled, deionized water (ddH_2O).
6. UV-transparent 96-well plates.
7. Transparent, flat-bottom polystyrene 1536-well plates.
8. Aluminum plate seals.

2.2 PSPL Screening

1. Purified kinase(s) to be assayed (*see* **Note 2**).
2. Kinase reaction buffer: If the optimal buffer is not known for the kinase, we generally use 50 mM HEPES, pH 7.4, 10 mM $MgCl_2$, 1 mM DTT, and 0.1 % Tween 20 (*see* **Note 3**).
3. 10 μCi/μL γ-[^{33}P]-ATP, 3000 Ci/mmol (*see* **Note 4**).
4. Streptavidin-coated membrane (SAM2 Biotin Capture Membrane, Promega), cut into 5.5″ × 1″ strips and stored at −20 °C.
5. Transparent, flat-bottom, polystyrene 1536-well plates.
6. Clear adhesive plate seals.
7. Whatman blotting paper, cut slightly larger than a multiwell plate, approximately 6″ × 4″.
8. 10 mM ATP, adjusted to pH 7.0 with 0.1 M NaOH (store at −20 °C).
9. Isopropanol.
10. 1 % SDS.
11. 0.1 % Tween 20.
12. Pin Cleaning Solution (V&P Scientific VP110).
13. SDS/TBS wash solution: 0.1 % SDS, 10 mM Tris–HCl, pH 7.5, 140 mM NaCl.
14. 2 M NaCl.
15. 2 M NaCl, 1 % H_3PO_4.

2.3 Equipment

1. UV-visible absorbance plate reader.
2. 1536 floating pin replicator (V&P Scientific), loaded with 240 (48 × 5 rows) 0.787 mm diameter 200 nL slot pins (V&P Scientific FP3S200). Place the 240 pins in five of the seven rows closest to the guide pins, leaving the two rows closest to the guide pins empty. This allows for the pin tool to access rows 2 through 6 of each quadrant of a 1536-well plate for the transfer of a complete peptide library (*see* Fig. 1 for quadrant layout of the 1536-well plate) (*see* **Note 5**).
3. 48-pin strip (also loaded with 0.787 mm diameter, 200 nL slot pins, V&P Scientific # FP3S200).

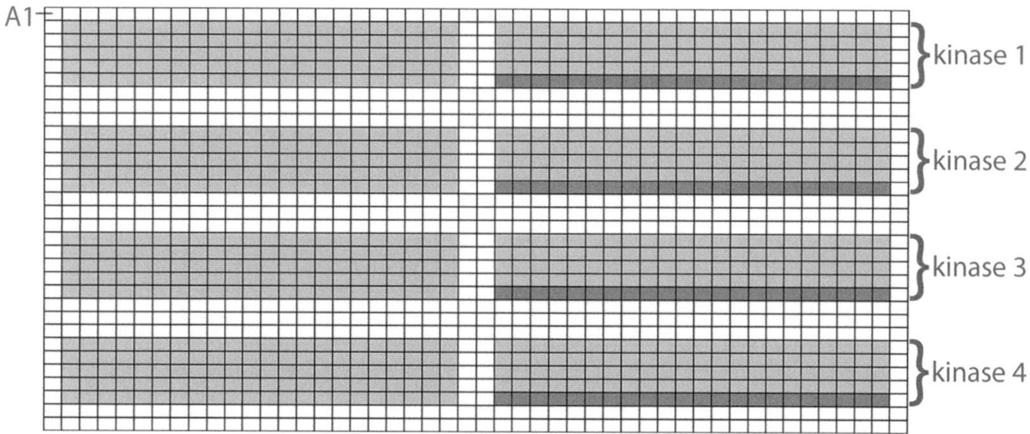

Fig. 1 Up to four kinases can be profiled per 1536-well plate. The *light gray boxes* represent wells in which peptide (laid out as in Fig. 3) and kinase are present; the *dark gray* represent wells in which only kinase/ATP are present and are used as background controls

4. Pin tool alignment frame with guide holes to allow the replicator loaded with 240 pins to access horizontal quadrants of a 1536-well plate (V&P Scientific, *see* Fig. 2A).
5. Pin tool alignment frame with guide holes to allow the 48-pin strip to access every row of a 1536-well plate (V&P Scientific, *see* Fig. 2B).
6. Several extra 0.787 mm diameter, 200 nL slot pins (type FP3S200).
7. Rubber mat (such as a 4″ × 6″ Speedball Speedy Carve Block).
8. Incubator set to 30 °C.
9. Low-speed refrigerated centrifuge with rotor capable of spinning multiwell plates at 4 °C.
10. Storage phosphor screen and cassette (Amersham Biosciences).
11. Phosphorimager (for example Molecular Imager FX Pro Plus, Bio-Rad) for scanning storage phosphor screen.

3 Methods

3.1 Preparation of PSPL Stock Plates

1. Purge a sufficient quantity of DMSO with oxygen by bubbling with a steady stream of argon for at least 5 min.
2. Dissolve powdered peptides in DMSO (15 μL/mg).
3. Dilute 1 μL of each DMSO stock to 500 μL with 20 mM HEPES, pH 7.4 in duplicate, and read the absorbance at 280 nm in a plate reader (a standard cuvette reading spectrometer may also be used).

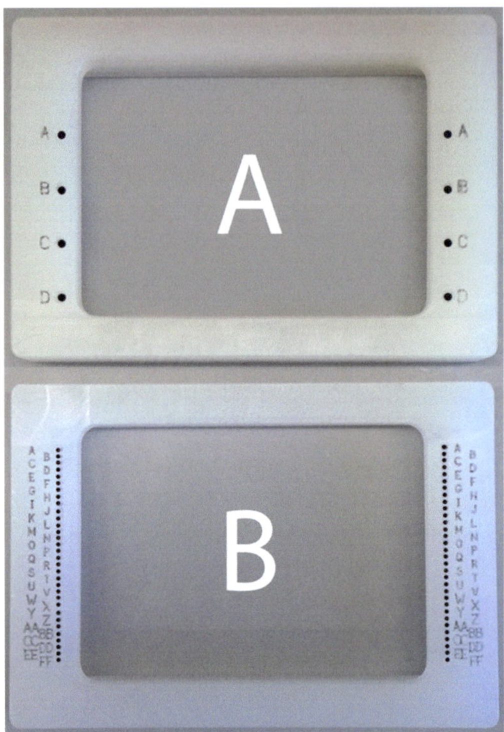

Fig. 2 Two pin tool alignment frames used in the procedure: The *A* frame has four holes for guide pins that allow a pin tool loaded with six rows of 48 pins to transfer the peptide library into one of the four quadrants of a 1536-well plate (*see* Fig. 1). The *B* frame has 32 pairs of holes to allow a 48-pin strip to transfer kinase/ATP to each row of a 1536-well plate

4. Back calculate the concentration of peptide in the DMSO solutions based on the molar absorptivity of the peptides (*see* **Note 6**).
5. Based on the absorbance readings, add additional DMSO to stock solutions to bring peptides to a concentration of 10 mM. Store DMSO stocks at −20 °C.
6. Dilute 1.5 μL of each 10 mM DMSO stock with 23.5 μL ddH_2O to generate working aqueous 0.6 mM stock solutions. Aqueous stocks may be stored at −20 °C.
7. Array 0.6 mM aqueous solutions in a 1536-well plates by adding 5 μL per well using the template shown in Fig. 3. Seal stock plates with aluminum seals and store at −20 °C. This plate can be used to assay approximately 16 kinases.

3.2 Peptide Library Screening

Both peptide stock and reaction plates should be kept cool on ice at all possible times to minimize evaporation (*see* **Note 7**). Up to four kinases can be profiled per 1536-well plate (Fig. 1). For the purposes of this protocol, we define four horizontal quadrants as

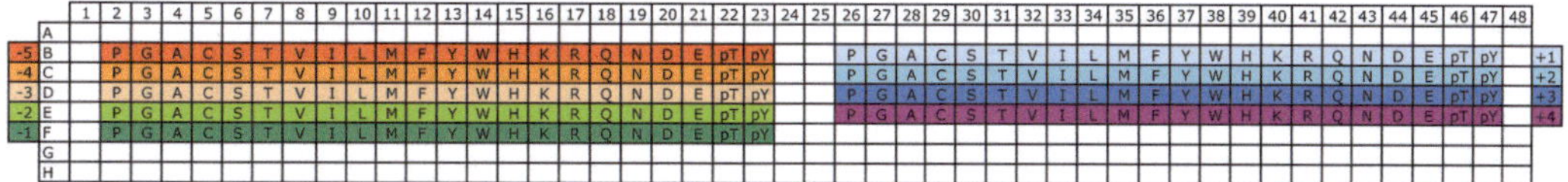

Fig. 3 Peptide library layout in stock plate and for kinase in the upper quadrant of reaction plate: The peptide sets with fixed residues N-terminal to the phosphosite (labeled −5 through −1) are on the left side of the plate (columns 2–23), and the peptide sets with fixed residues C-terminal to the phosphosite (labeled +1 through +4) are on the right side of the plate (columns 26–47). *Letters* indicate the fixed amino acid, where pT is phosphothreonine, and pY is phosphotyrosine

Quadrant I = rows A–H, Quadrant II = rows I–P, Quadrant III = rows Q–X, and Quadrant IV = rows Y–FF.

1. Thaw the peptide library stock plate on the benchtop. Shake or vortex the plate to mix, chill it on ice, spin plate in the refrigerated centrifuge (1400 × *g* for 1 min), and return to ice.
2. Chill a 1536-well plate to be used for the kinase reactions. Aliquot 2 μL of ice-cold kinase reaction buffer into each well of the first seven rows of each quadrant, leaving the eighth row of the quadrant empty (*see* **Note 8**).
3. Centrifuge the reaction plate (1400 × *g* for 1 min) to remove air bubbles and bring the solution to the bottom of the wells.
4. Unseal the peptide stock plate, and place on the benchtop alongside the reaction plate. Place alignment frames with quadrant guide holes over both plates (Fig. 2A).
5. Dip the pins of the 240-pin replicator in 0.1 % Tween, and then blot out the liquid on Whatman paper.
6. Align the replicator with the upper quadrant of the peptide stock plate by placing the guide pins into the appropriate guide holes. Insert the pins into the wells to transfer aliquots of each peptide solution to the pins. Dip the pins in the stock plate five times moving the device up and down, breaking the surface of the liquid each time (*see* **Note 9**).
7. Transfer the peptides from the replicator to the first quadrant of the reaction plate by placing the guide pins into the appropriate guide holes as above, and dipping the pins into the wells. Move the device up and down five times to mix thoroughly.
8. Blot out excess liquid from the pin tool on Whatman paper.
9. If more than one kinase is to be analyzed, repeat **steps 5–8** to use additional quadrants of the reaction plate.
10. Wash the pins by dipping in the following solutions five times, blotting out the liquid in between each solution: one wash with 0.1 % Tween, two washes with ddH_2O, and a final wash with isopropanol. Blot out isopropanol and allow the pin tool to air-dry.

11. Cover the stock plate with a fresh aluminum seal and return to storage at –20 °C.
12. Prepare 120 μL 10× kinase/ATP solutions containing the desired amount of each kinase (*see* **Note 10**), 550 μM cold ATP, and 0.33 μCi/μL γ-[^{33}P]-ATP in kinase reaction buffer.
13. Place the 32-row alignment frame (Fig. 2B) on the reaction plate.
14. Add 10× kinase/ATP solution to a disposable reservoir.
15. Transfer the 10× kinase/ATP solution to the reaction wells of the each row of the plate that contains peptide solutions as follows. First prime the 48-pin strip by dipping in 0.1 % Tween and blotting out the liquid. Next, dip the pins into the 10× kinase/ATP solution, ensuring that the pins are coated with solution. Using the alignment holes on the sides of the frame, place the pins into the second row of the appropriate quadrant (row B for the first quadrant). Dip the pins in the buffer five times by moving the strip up and down to ensure complete transfer of the solution to the wells. In between rows, blot out excess liquid on Whatman paper (caution: blotting paper will be radioactive), and wash pin strip twice with 0.1 % Tween and four times with ddH_2O by dipping the pin strip in each solution five times and blotting in between each wash. Repeat this step for the four other rows in the quadrant.
16. Wash the 48-pin strip by soaking the pins in 1 % SDS for 2 min. Blot out excess liquid, and then wash twice with ddH_2O and once with isopropanol. Blot out isopropanol and allow pins to air-dry.
17. Repeat **steps 14–16** to add any additional kinases to the remaining quadrants (*see* Fig. 1).
18. Seal the reaction plate with plastic adhesive tape and incubate for 2 h at 30 °C (*see* **Note 11**). During the incubation, allow streptavidin membrane strips to warm to room temperature.
19. When the incubation is complete, chill reaction plate on ice, remove the seal, and place the four-quadrant alignment frame (Fig. 2A) over the plate.
20. Tape the streptavidin membrane (labeled with a pencil if profiling more than one kinase) onto the rubber mat.
21. Prime the 240-pin replicator by dipping the pins into 0.1 % Tween and blotting out the solution on Whatman paper.
22. Transfer aliquots from one reaction quadrant to the replicator pins by dipping the pins into the reaction plate, using the guide holes on the alignment frame. Raise and lower the pins five times to ensure efficient transfer.

23. In one even motion, lower the pins onto the streptavidin membrane. Hold the pins on the membrane for a few seconds, rocking the pin tool back and forth slightly.
24. Lift the pin tool from the membrane and inspect to confirm that all aliquots were transferred successfully. If any aliquots failed to transfer, use an individual pin to manually transfer aliquots from the reaction plate.
25. After 20 s, remove the streptavidin membrane from the rubber mat and submerge it in the first wash solution (0.1 % SDS/TBS) to quench the reactions.
26. Wash the pin device by dipping five times into and blotting between the following solutions: once in 0.1 % Tween, twice in ddH_2O, and once in isopropanol. Blot out isopropanol and allow the device to air-dry.
27. Repeat **steps 20–26** for each of the remaining quadrants.
28. Swirl the membranes in the first wash solution a few times, and then dispose of the solution as radioactive liquid waste according to the protocols of your institution.
29. Wash membranes twice with 100 mL 0.1 % SDS/TBS, twice with 100 mL 2 M NaCl, and twice with 2 M NaCl/1 % H_3PO_4, 3 min per wash. Rinse the membranes twice briefly with dH_2O to remove excess buffer. Dispose of the wash solutions as appropriate for mildly (<1 μCi) radioactive liquid waste.
30. Allow membranes to air-dry on a piece of aluminum foil for approximately 30 min, wrap in saran wrap, and expose to a storage phosphor screen overnight.
31. The following day, scan the phosphor screen with the phosphor imager. If the signal is weak, a longer exposure (generally not more than 1 week) may be required.

3.3 Data Analysis

1. The image will appear as an array of spots (*see* Fig. 4a). Quantify the spot intensities using the image analysis software accompanying the phosphor imager. Export raw spot intensity data into a spreadsheet.
2. The wells in the sixth row, columns 26–47 of each quadrant (dark gray in Fig. 1), contain kinase and radiolabeled ATP but no peptide, and provide the background signal. Average the signal from these positions, and subtract this average background value from all the spot intensities to be quantified.
3. Normalize the data for each spot by dividing its background-subtracted signal from the average value for all the spots in the same position. Once normalized in this way, the average value at each position is 1. Positively selected residues have values greater than 1, and negatively selected residues have values less than 1 (*see* **Note 12**).

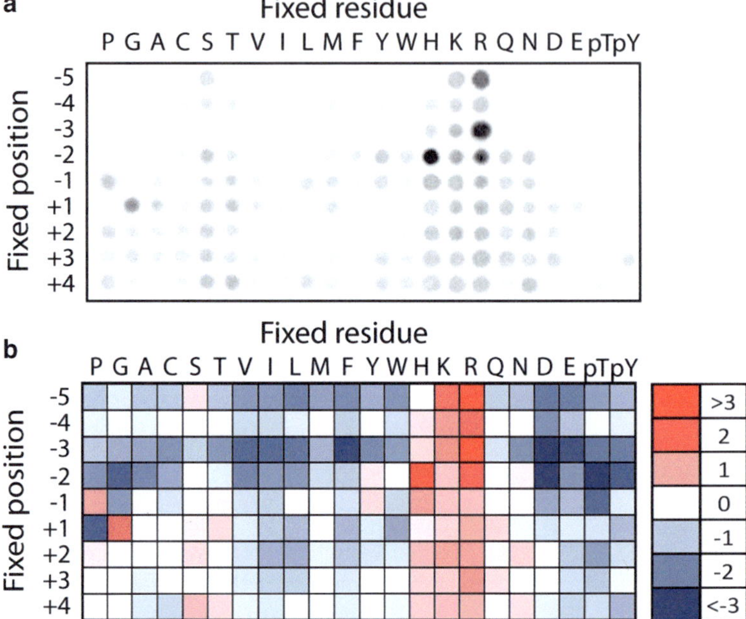

Fig. 4 Representative results from PSPL assay: (**a**) Raw autoradiogram showing array of radioactive spots for the kinase Pim3. Spots arising from wells on the left and right sides of the plate, respectively, were arranged using Adobe Photoshop so that the peptide positions are presented in ascending order. (**b**) Normalized spot intensities from panel **a** were converted into a heat map showing positively (*red*) and negatively (*blue*) selected residues at each position

4 Notes

1. For analysis of tyrosine kinases, we have used a similar peptide library with the general sequence G-A-X-X-X-X-X-Y-X-X-X-X-A-G-K-K(biotin), in which the degenerate X positions are comprised of all the amino acids except Tyr and Cys. The 240-pin tool used to transfer the libraries should be washed if the library (e.g., tyrosine vs. serine/threonine library) is changed between runs. To clean the pin tool, allow it to soak in pin cleaner for 2 min, and then wash twice with ddH_2O and once with isopropanol. Allow the pin tool to air-dry before continuing with the protocol.

2. A suitable expression system for producing active kinase is required. If possible, bacterial expression systems are preferred because there is little concern that contaminating endogenous kinases will produce a signal on the peptide arrays. However, most kinases are inactive or insoluble when produced in bacteria, necessitating the use of eukaryotic cell (insect, mammalian, or yeast) expression systems. These expression systems always

carry the risk of having contaminating kinases produce a signal that could be spuriously attributed to the kinase of interest. We have encountered cases where kinases that appear to be highly pure on a Coomassie-stained polyacrylamide gel have contaminating kinase activities present. We recommend for kinases produced in eukaryotic expression systems that a kinase-inactive mutant also be produced and subjected to PSPL screening. We tend to avoid using commercial sources of kinase because this control is generally unavailable. We have encountered cAMP-dependent protein kinase (PKA) as a common contaminant from mammalian cells. PKA contamination can be managed by including the specific peptide inhibitor PKI (100 nM) in the reactions.

3. In our hands, this buffer works for most kinases. If the kinase is not sufficiently active, substituting $MnCl_2$ for $MgCl_2$ will activate some kinases. The assay is compatible with a wide range of buffer conditions. If a different buffer is being used, it is important to add a low concentration of a nonionic detergent (i.e., 0.1 % Tween 20) so that the pin tools function properly.
4. γ-[^{32}P]-ATP can also be used if appropriate safety precautions are taken. We recommend γ-[^{33}P]-ATP because it does not require a safety shield.
5. Though the procedure can be performed with a single replicator, having two replicators available is useful. One is used for transferring peptides from the stock plates to the reaction plates. While the other, dedicated for radioactive use, is used for transferring aliquots of the reactions to the streptavidin membrane. Having two replicators allows many steps of the procedure to be performed in laboratory spaces not approved for radioactive use.
6. Because of variable solvent content, peptide weight is not a reliable parameter for determining the concentration of peptide solutions. It is preferable to use UV absorbance based on the molar extinction coefficients of Trp (5690 M^{-1} cm^{-1}) and Tyr (1200 M^{-1} cm^{-1}) in peptides. We assume that the mixture positions are 5.9 % Trp and 5.9 % Tyr, which provides a molar absorptivity value of 4560 M^{-1} cm^{-1} for most peptides, 10,250 M^{-1} cm^{-1} for peptides with fixed Trp residues, 5840 M^{-1} cm^{-1} for peptides with fixed Tyr residues, and 5010 M^{-1} cm^{-1} for peptides with fixed phosphoTyr residues.
7. We typically place custom-machined aluminum blocks ($11.5 \times 7.2 \times 0.4$ cm) underneath the multiwell plates to facilitate uniform cooling on ice. By placing the blocks on top of a larger aluminum plate ($21 \times 15 \times 0.4$ cm), all liquid transfer steps can be performed within a tray of ice rather than on the

benchtop. This is useful for minimizing evaporation, particularly if multiple kinases will be screened at one time. If these are not available, re-chilling the plates by placing on ice in between transfers is important to reduce evaporation.

8. By adding buffer to all wells in rows 1 through 7, reaction wells containing peptide and kinase are surrounded by "blank" wells containing only buffer (*see* Figs. 1 and 3), which helps to prevent evaporation in the reaction wells.
9. The slot pins in the replicator withdraw aliquots of liquid by capillary action. Occasionally a pin will fail to pick up liquid, resulting in no liquid transfer. Dipping the pins into the liquid multiple times during transfers helps to ensure that all pins are filled. When dispensing liquid into the destination plate, dipping the pins repeatedly serves to mix the contents of the pins with buffer in the well, ensuring complete transfer of the contents.
10. How much kinase is required for the PSPL procedure depends on multiple factors, including the level of activity of the kinase preparation and the intrinsic ability of the kinase to phosphorylate peptides, which can vary considerably from kinase to kinase. In our experience the amount of kinase used in this protocol has ranged from 240 ng to 100 μg. We typically start with 10 μg of kinase (providing approximately 7.5 ng/μL kinase in the final reaction wells) and adjust up or down depending on the results of an initial PSPL assay.
11. We use a dedicated multiwell plate incubator to keep reaction plates at 30 °C. If a plate incubator is not available, an alternative is to line a Tupperware container with wet paper towels to make a humidified chamber that can be placed inside a standard 30 °C incubator.
12. One way to visualize the phosphorylation site motif is to display the data as a heat map (*see* Fig. 4b). Heat maps can be generated using Microsoft Excel. First log transform the normalized data so that the average value at a single position is 0. Then use the conditional formatting feature (three-color scale) to shade the negative, zero, and positive values with different colors.

Acknowledgments

The authors are supported by NIH grants R01 GM102262 and R01 GM104047.

References

1. Ubersax JA, Ferrell JE Jr (2007) Mechanisms of specificity in protein phosphorylation. Nat Rev Mol Cell Biol 8:530–541
2. Pinna LA, Ruzzene M (1996) How do protein kinases recognize their substrates? Biochim Biophys Acta 1314:191–225
3. Kennelly PJ, Krebs EG (1991) Consensus sequences as substrate specificity determinants for protein kinases and protein phosphatases. J Biol Chem 266:15555–15558
4. Yaffe MB, Leparc GG, Lai J, Obata T, Volinia S, Cantley LC (2001) A motif-based profile scanning approach for genome-wide prediction of signaling pathways. Nat Biotechnol 19: 348–353
5. Songyang Z, Blechner S, Hoagland N, Hoekstra MF, Piwnica-Worms H, Cantley LC (1994) Use of an oriented peptide library to determine the optimal substrates of protein kinases. Curr Biol 4:973–982
6. Meggio F, Pinna LA (2003) One-thousand-and-one substrates of protein kinase CK2? FASEB J 17:349–368
7. Shabb JB (2001) Physiological substrates of cAMP-dependent protein kinase. Chem Rev 101:2381–2411
8. Zhu G, Fujii K, Belkina N, Liu Y, James M, Herrero J et al (2005) Exceptional disfavor for proline at the P+1 position among AGC and CAMK kinases establishes reciprocal specificity between them and the proline-directed kinases. J Biol Chem 280:10743–10748
9. Alexander J, Lim D, Joughin BA, Hegemann B, Hutchins JR, Ehrenberger T et al (2011) Spatial exclusivity combined with positive and negative selection of phosphorylation motifs is the basis for context-dependent mitotic signaling. Sci Signal 4:ra42
10. Fujii K, Zhu G, Liu Y, Hallam J, Chen L, Herrero J et al (2004) Kinase peptide specificity: improved determination and relevance to protein phosphorylation. Proc Natl Acad Sci U S A 101:13744–13749
11. Papinski D, Schuschnig M, Reiter W, Wilhelm L, Barnes CA, Maiolica A et al (2014) Early steps in autophagy depend on direct phosphorylation of Atg9 by the Atg1 kinase. Mol Cell 53:471–483
12. Lipchik AM, Killins RL, Geahlen RL, Parker LL (2012) A peptide-based biosensor assay to detect intracellular Syk kinase activation and inhibition. Biochemistry 51:7515–7524
13. Hutti JE, Porter MA, Cheely AW, Cantley LC, Wang X, Kireev D et al (2012) Development of a high-throughput assay for identifying inhibitors of TBK1 and IKKepsilon. PLoS One 7:e41494
14. Zhang J, Allen MD (2007) FRET-based biosensors for protein kinases: illuminating the kinome. Mol Biosyst 3:759–765
15. Miller ML, Jensen LJ, Diella F, Jorgensen C, Tinti M, Li L et al (2008) Linear motif atlas for phosphorylation-dependent signaling. Sci Signal 1:ra2
16. Linding R, Juhl Jensen L, Pasculescu A, Olhovsky M, Colwill K, Bork P et al (2008) NetworKIN: a resource for exploring cellular phosphorylation networks. Nucleic Acids Res 36(Database issue):D695–D699
17. Lam HY, Kim PM, Mok J, Tonikian R, Sidhu SS, Turk BE et al (2010) MOTIPS: automated motif analysis for predicting targets of modular protein domains. BMC Bioinformatics 11:243
18. Songyang Z, Carraway KL III, Eck MJ, Harrison SC, Feldman RA, Mohammadi M et al (1995) Catalytic specificity of protein-tyrosine kinases is critical for selective signalling. Nature 373:536–539
19. Tegge WJ, Frank R (1998) Analysis of protein kinase substrate specificity by the use of peptide libraries on cellulose paper (SPOT-method). Methods Mol Biol 87:99–106
20. Kettenbach AN, Wang T, Faherty BK, Madden DR, Knapp S, Bailey-Kellogg C et al (2012) Rapid determination of multiple linear kinase substrate motifs by mass spectrometry. Chem Biol 19:608–618
21. Douglass J, Gunaratne R, Bradford D, Saeed F, Hoffert JD, Steinbach PJ et al (2012) Identifying protein kinase target preferences using mass spectrometry. Am J Physiol Cell Physiol 303:C715–C727
22. Trinh TB, Xiao Q, Pei D (2013) Profiling the substrate specificity of protein kinases by on-bead screening of peptide libraries. Biochemistry 52:5645–5655
23. Mok J, Kim PM, Lam HY, Piccirillo S, Zhou X, Jeschke GR et al (2010) Deciphering protein kinase specificity through large-scale analysis of yeast phosphorylation site motifs. Sci Signal 3:ra12
24. Hutti JE, Jarrell ET, Chang JD, Abbott DW, Storz P, Toker A et al (2004) A rapid method for determining protein kinase phosphorylation specificity. Nat Methods 1:27–29
25. Kemp BE, Graves DJ, Benjamini E, Krebs EG (1977) Role of multiple basic residues in determining the substrate specificity of cyclic AMP-dependent protein kinase. J Biol Chem 252: 4888–4894
26. Alessi DR, Caudwell FB, Andjelkovic M, Hemmings BA, Cohen P (1996) Molecular

basis for the substrate specificity of protein kinase B; comparison with MAPKAP kinase-1 and p70 S6 kinase. FEBS Lett 399:333–338

27. Dale S, Wilson WA, Edelman AM, Hardie DG (1995) Similar substrate recognition motifs for mammalian AMP-activated protein kinase, higher plant HMG-CoA reductase kinase-A, yeast SNF1, and mammalian calmodulin-dependent protein kinase I. FEBS Lett 361:191–195
28. Gwinn DM, Shackelford DB, Egan DF, Mihaylova MM, Mery A, Vasquez DS et al (2008) AMPK phosphorylation of raptor mediates a metabolic checkpoint. Mol Cell 30:214–226
29. Gonzalez FA, Raden DL, Davis RJ (1991) Identification of substrate recognition determinants for human ERK1 and ERK2 protein kinases. J Biol Chem 266:22159–22163
30. Flotow H, Graves PR, Wang AQ, Fiol CJ, Roeske RW, Roach PJ (1990) Phosphate groups as substrate determinants for casein kinase I action. J Biol Chem 265:14264–14269
31. Songyang Z, Lu KP, Kwon YT, Tsai LH, Filhol O, Cochet C et al (1996) A structural basis for substrate specificities of protein Ser/Thr kinases: primary sequence preference of casein kinases I and II, NIMA, phosphorylase kinase, calmodulin-dependent kinase II, CDK5, and Erk1. Mol Cell Biol 16:6486–6493

Index

Hicham Zegzouti and Said A. Goueli (eds.), *Kinase Screening and Profiling: Methods and Protocols*, Methods in Molecular Biology, vol. 1360, DOI 10.1007/978-1-4939-3073-9, © Springer Science+Business Media New York 2016

MIX
Papier aus verantwortungsvollen Quellen
Paper from responsible sources
FSC® C105338

If you have any concerns about our products,
you can contact us on
ProductSafety@springernature.com

In case Publisher is established outside the EU,
the EU authorized representative is:
Springer Nature Customer Service Center GmbH
Europaplatz 3, 69115 Heidelberg, Germany

Printed by Libri Plureos GmbH
in Hamburg, Germany